ARGUMENTS

OF

WILLIAM W. HUBBELL, ESQ.,

OF PHILADELPHIA,

ON BEHALF OF THE DEFENDANTS,

BEFORE

HON. SAMUEL NELSON,

IN THE CASE OF

ROSS WINANS vs. ORSAMUS EATON, et al.,

IN THE

CIRCUIT COURT OF THE UNITED STATES OF THE NORTHERN DISTRICT OF NEW-YORK, AGAINST A MOTION FOR AN INJUNCTION TO RESTRAIN THE DEFENDANTS FROM CONSTRUCTING AND VENDING THE RAILROAD CARS COMMONLY KNOWN AS THE

"EIGHT-WHEEL RAILROAD CAR,"

ALLEGED TO INFRINGE ROSS WINANS' PATENT OF OCT. 1, 1834.

ALBANY:
WEED, PARSONS & CO., PRINTERS.
1853.

ARGUMENT

OF

WILLIAM W. HUBBELL, ESQ.

MONDAY, *August* 15*th.*

WM. W. HUBBELL, Esq., said:

If the Court please:

The opening argument of one of the opposing counsel on behalf of the plaintiff, was commenced with allusions to smoke, gunpowder, and the morale of the case; and, after consuming a large quantity of the villainous salt-petre, and raising considerable smoke, and after discharging the last shot in his locker; upon looking round, we find that no one is killed—no one is wounded. And we are inclined to believe, that the obscurity attempted to be thrown over this case, has been with a view to conceal the real issues involved, and to mislead the judicial mind; notwithstanding which, however, we hope to be able to make the merits appear fully and clearly in favor of the defendants.

Mechanics, when properly viewed, are simple, when looked into, without an intention to pervert; and men who have lived before us, were not all void of the commonest principles of common sense, as this court is asked to believe by the plaintiff's attorney, in order that he may be allowed to levy an enormous tax, on a thing which he neither describes nor claims, and which looks back, with unerring confidence, to the earliest part of the present century, as the point of time on which it rests, and whose existence here is coeval with the railroads of this country: whose growth has been with their growth, still retaining the same vitality, and living in the same substantial form.

Remarks were also made with reference to the history of railroads, but strange to relate, those remarks were not in accordance with that history.

It is shown by competent testimony in this case, that railways in the interior of England, at a very early day, were very much curved; that was their general character. It also appears that the Stockton and Darlington road, was one of the first in existence, even before the Liverpool and Manchester road, and is referred to in Tredgold, p. 20, in these words:

"And an extensive railway from Stockton by Darlington to the collieries on the southwest side of the county of Durham, is now nearly completed. It proceeds from Stockton in a westerly direction, and about 3½ miles from thence, a branch to the south of 2 miles, leads off to Yarm; the main line passes close to Darlington, and about 4 miles beyond Darlington, a branch to the south, of nearly 2 miles, leads to Pierce Bridge. About 5 miles further on the line, the Black Boy branch leads off in a northeasterly direction to the Black Boy and Coundon collieries. The extent of this branch is upwards of 5 miles. The main line continues past Evenwood to near the Norwood collieries and returns in a northeasterly direction to the Etherly and Wilton-Park collieries. The total extent of the main line is about 32 miles. It is formed with edge rails; and in the last Act of Parliament for this railway, there are clauses to enable the company to make use of locomotive engines."

The Liverpool and Manchester road was afterwards built between two large cities, and, from that communication, necessarily, passenger travel, as well as the transportation of freight, was attracted to that road. A large business was done on it; yet that fact is entirely disconnected from the subject of an 8 wheel car; for, on that road, four wheel cars are used at the present day. It does not necessarily follow, that where persons are conveyed on any road, that the 8 wheel car must be used, for it is notorious that the four wheel cars are used, as I have said, and generally, on the English roads, at the present day.

The 8 wheel car is not suited to the disposition and habits of the English people. *It* is republican in its character, because it allows persons to mingle freely among each other, while the four wheel car admits of exclusive parties of the aristocracy of that country, and is more consistent with the general disposition and mode of living. These are the principal reasons why four wheel cars are used in England, and eight wheel cars in the United States.

The Liverpool and Manchester road is proven to have had considerable curvatures. Mr. Thompson's dep., No. 3, p. 1, fols. 2, 3, testifies "that it has considerable curvatures:" and yet these four wheel cars run over it.

The counsel also adverted to the locomotive steam engine, and said that the idea of adhesion of the wheels did not exist prior to 1830; and, as authority upon this point, read from a book published in that year. And, while reading, it is not a little singular that this passage occurred, in substance : " *The wheels slipped when snow was on the track*," which shows at once that the wheels propelled by adhesion to the rails. They were not cog-wheels, and the friction, or adhesion, was obviated by snow upon the track.

It was not only known then, prior to 1830, as shown by the book from which he read, but we find it existing, in actual use, long prior to 1830, in numerous places.

This is shown in Wood's Treatise in 1825, in the engine called the Wood engine.

This model here, they have presented as their own illustration of that engine, and it is shown to move by adhesion to the rail, which is conclusive that this power of traction by the resistance of adhesion to the rails, was known at that period, 1825.

An entire train is shown in fig. 7, plate 2, fully demonstrating that fact.

In Tredgold's Treatise of 1825, pages 74 and 176, this is also set forth.

The practice of using locomotives to draw by adhesion to the rails is very old, and there is no novelty in it; neither was it by any means new in 1830.

Cogs, and endless chains or ropes, also spoken of, are used on Marine railways now, which proves that there was no novelty in them. They were all old modes of applying the steam power.

There was an important feature which gave the locomotive the impetus it now possesses, discovered at an early day; this was in the method of generating steam. We see in Tredgold, of 1825, that the locomotive had a very tall chimney, sometimes from 12 to 13 feet high; and they obtained draft, at that period, by the natural action of the air and gases in the chimney.

It was discovered by an engineer on the Stockton and Darlington road, that by inserting the exhaust pipes into the chimney and contracting their mouths so as to throw the exhaust steam with velocity up the chimney, it would drive out the air and create a partial vacuum below, and thus cause a powerful draft to rush through the furnace.

Then it was that the speed increased, and without that one feature, the locomotive of the present day would not be able to run more than 8 or 10 miles an hour. The impetus obtained was by means of generating steam by exhaustion, as stated, and not by the adhesion of the wheels to the rails. In this country, steam was successfully introduced on the Charleston road in 1830, as shown by the testimony of Allen and Detmoold. On the Newcastle road in 1832, as shown by Dorsey. On the Mohawk and Hudson road in 1832, as shown by Whitney, Mathews and Jervis. On the Baltimore and Ohio road they were experimenting from 1831 to 1834; and in 1834–5 they had engines of considerable promise. The engines of 1831–2 were a failure. The testimony of Mr.

Dorsey shows, that on a trial of speed for a considerable distance, horses had beat the best locomotive they had produced. His affidavit to that effect, is to be found in defendants' No. 1, p. 78, fol. 296.

The "Arabia," in 1834, which killed Phineas Davis on its trial trip, was the first comparatively successful engine they had. The fact is mentioned, I think, in Mr. Gillingham's 9th annual report, A. D. 1835, ps. 11–12, and dated 1st. Oct., of that year; also in the affidavit of Edward Gillingham, in defendant's No. 4, p. 56, fol. 215.

Such is a general outline of the history of the locomotive, and it has been increasing in weight and speed, and cars have also been increasing in size and number, in accordance with the amount of business transacted on railroads, and as railroads themselves have increased.

My next subject is:

That this suit, with others, are a surprise at this late day in the State of New-York and in the New England states.

Suits of this character being suits in equity praying for injunctions, have been brought not only against the present defendants, but against several railroad companies: the "Albany and Schenectady Railroad Company," the "Utica and Schenectady Railroad Company," the "Syracuse and Utica Railroad Company," the Rochester and Syracuse Railroad Company," the "Buffalo and Rochester" (now the Central road.)

Against all these companies injunctions have not yet been moved for.

Also, since the bringing of these suits, suit has been brought against the Eastern Railroad Company, in Massachusetts, which has gone on to final hearing without a motion for an injunction. This suit is now in course of preparation for final hearing; and is expected to come up some time in November, or about that time.

The order of the court is, that the proofs shall be in by the 7th Oct., but they are so very voluminous that its practicability is somewhat uncertain.

These suits were preceded by one against a small road, the "Troy and Schenectady," and through that, and the others pending in this court and in Massachusetts, the claims and demands of Ross Winans, which followed it, by the circular of Gould and Spencer, in 1852–53, have become known to car builders and railroad companies in the northern states eighteen or nineteen years after the date of the patent. In fact, these suits are a speculation by third parties, in proof of which, I will read from No. 3 (defendants' proof), p. 160, folio 638–9; 640, and p. 162, folio 648–9. The first, is the affidavit of Charles Minot. He says:

"I have known Ross Winans, personally, for several years, and have had very frequent conversation with him on railroad matters generally. In some of my conversations with him of late years, he stated that he had a great many patents, and never realized much from them, and that he was going to take a new course in future; that he was not going to litigate them at his expense. And in conversation with regard to his eight-wheeled car patent, he gave me to understand that he was not much interested in the suits under it; and that he had agreed with other parties at Albany, that they should prosecute the suits, and if there was nothing gained or got out of it, that he, Winans, was to be at no expense; that the other parties were to pay the expense, and if anything was made out of that, they were to have a share. After this I received a circular from Charles D. Gould of Albany, and J. A. Spencer, of Utica, claiming the right to compensation for the using of eight-wheeled cars, on the New-York and Erie Railroad, alleging that they were an infringement of Ross Winans' patent, of Oct. 1st, 1834. This was after the termination of the suit of Ross Winans *vs.* the Troy and Schenectady road. After receiving the circular from Mr. Gould and Mr. Spencer, I talked with Mr. Ross Winans about it, and he said that Gould was managing that matter, and was interested in it. Winans did not urge any settlement. I was conversing with him about some engines of his construction; and he said nothing further about the eight-wheeled car, than I have stated.

"I annex the said circular hereto, marked A. I."

This circular, after alleging that a patent had been granted to Ross Winans, the 1st Oct., 1834, and after setting out the Troy and Schenectady case, goes on to state in folio 649:

"This letter is intended to bring this subject to the notice and attention of the different railroad companies in the United States, which have not already purchased the right to use Mr. Winans' 'Eight-Wheel Railroad Car,' and to give information that the undersigned are fully authorized, and are ready, to settle and adjust all matters in relation to the past use of the car, and to convey the right to its future use, during the continuance of the patent, on the most liberal terms; and the hope and expectation are indulged, that the several railroad companies will promptly avail themselves of the opportunity so to do.—Dated 15th September, 1851.

"CHARLES D. GOULD, *Albany*,
"J. A. SPENCER, *Utica, Attorney and Counsel for* Ross Winans.

This is signed Charles D. Gould, not as the agent of Winans, but absolutely in his own right, with an assertion that he has power to convey that right, and that he is "fully authorized and ready to settle and adjust all matters in relation to the past use of the car." This statement, in connection with that of Winans, shows that this is a speculation, undertaken by Mr. Gould, on condition that he shall pay all expenses, and if he be successful, Winans and he will divide the spoils.

This speculation has been planned with a view to approach that basis, which equity requires to sustain a bill; by first attaching small firms and corporations—little able to make a defence—accompanied by offers of settlement for small sums, to induce them, if possible, to submit to small but unjust exactions, rather than incur heavy expenses in litigation. There is strong policy in this, and it is resorted to for the purpose—

at this very late day—of establishing some foundation in equity for a dormant patent, a patent unheard of and unknown in the New England states, until fourteen or fifteen years after it was issued.

This policy is followed, evidently, with the view of approaching the large corporations, or heavy railroad interests, from whom princely fortunes are demanded, backed by suits, and threats, to stop all their cars and business by injunction, if not complied with.

In the case before Judge Conkling, at Auburn, $100,000 was stated as the demand from the Central Railroad line. This was stated by the counsel for the plaintiff in July of last year.

The suit against the Troy and Schenectady road was a small affair, in comparison with the present.

It was a surprise on that road. It was struggling for an existence; and its current expenses exceeded its receipts, as shown by the official state of the reports, and was therefore little able to defend itself for want of means, and has since been sold out for less than its original cost.

Judge NELSON. It was represented by as able counsel as could have been employed.

Mr. HUBBELL. The counsel admit that they knew but little of the mechanics in that case, as the affidavit of Mr. Buel, in dep. No. 2, p. 4, fol. 16, will show.

Judge NELSON. It was tried by Mr. Buel and Mr. Stevens, and Mr. Stevens is one of the ablest gentlemen in his profession.

Mr. HUBBELL. That it was tried by able counsel, I have no doubt; but, the history of the eight-wheel car had not been a study of the profession at that time. The defences before this court were, consequently, not then known; and a knowledge of defences in patent cases, is obtained by perseverance, among strangers, in unknown places, and surrounded by difficulties.

In addition to the difficulties, which always surround investigations such as these, it appears, by the testimony of Mr. Atkinson, treasurer of the Baltimore and Ohio Railroad Company, that the papers which would convey information respecting that subject, were obtained by Winans, in September, 1852, who promised, at the same time, to return them in sixty days, and which, notwithstanding, are still in Winans' possession, and have not been returned.

From the date of the existence of the Quincey car till the present time, many obstacles have occurred, and death has removed beyond our reach many of our principal witnesses. Since the Troy and Schenectady case, one—Mr. Aler—is dead; and so they are, one by one, taken away.

Every lawyer, if he has had experience in patent cases, and happens to have been on the part of the defence, knows these difficulties, in contending against *old* and *dormant* patents. And I presume that your Honor's experience has taught the same truths. Not only has the patent remained *dormant*, but Ross Winans himself has been engaged in pursuits so widely different from that of a practical car-builder, that the remotest pretension of his being, or considering himself connected with the origin, construction, or use of the eight-wheel car, was not entertained by the defendants, nor the railroads, in this part of the country, for which they built cars. He was known only as a builder of locomotives and wheels on the Phineas Davis' plan, and was only notorious for planning and building locomotives, in which business he, for many years, has been and is now largely engaged. And also, he was notorious as having claimed, under a patent, the outside bearings on railroad carriages, which, in the suit of Winans *vs.* The Boston and Providence Railroad Company, was proved not to be his invention, and his suit failed, [See 2d Story, C. C. Rep., 412.]

That was the termination of that patent, and nothing more has been heard of it. There was no surrender—no re-issue: the whole thing was abandoned.

The engines were known to be used almost exclusively on the Baltimore and Ohio road, running at a speed of from ten to twelve miles an hour in the transportation of freight, while other roads required better planned and constructed, more economical and efficient engines, on entirely different arrangements of machinery, and of far superior speed. His present position is, therefore, a surprise upon the defendants, and the Northern railroads and car builders, calculated to find them but little prepared to meet a dormant patent, speculatively sprung upon them after a sleep of 18 or 19 years, in relation to a car, the history of which, on the Baltimore and Ohio road, is peculiar, and the general history of which runs back in this country to 1829, and in England and Ireland, its construction and principles are found described in 1825, and back as far as 1814. There was no cause, therefore, to suppose that any one claimed a patent for an eight-wheel car. The idea was never entertained that any one claimed a patent, and much less Winans, who had full opportunity to set forth his claim, if he thought that the cars used by railroads infringed his patent, and notwithstanding which opportunity, he never made any such pretension.

Lawyers called upon to defend such cases, might well be surprised and non-informed as to the true principles and history of the car, let them be the ablest that the country affords.

Long years of practice and customary use, had settled the principles in the construction of the car.

Inquiry as to the embodiment of that construction and principle in language for the use of courts and juries, was neither afloat nor in requisition. There were no demands made to set scientific and legal men upon the inquiry.

There was no cause to investigate the matter in such a light, and to embody it in such language, as to be able to present it clearly to the mind of a court or to a jury. It appeared to be a public right; a right immemorial almost, in its character, and one which no person supposed was questioned in the slightest degree. Under such circumstances, I say that the Troy and Schenectady suit was a surprise, and this suit is a surprise, in which, however, the plaintiff is on his *third* application for a preliminary injunction, he having failed on the two former motions; but finding *us*, on this much better prepared for the contest than we were on either of the two former, in both of which he was DEFEATED. Then we knew but little of the history of this matter, for it was deemed unnecessary, prior to that time, to investigate it fully.

When a man undertakes to investigate a subject of this kind, however powerful in intellect, or skilled he may be in points of law, he must have time—time not only to examine it in a scientific point of view, but time to go over this entire continent, if necessary, to ascertain, from one here, and another there, who have been scattered since 1829, what they know respectively about the matter.

In the next place, the position of the defendants in this case is peculiar. They are men of high business standing and responsibility, who by years of industry and enterprise have amassed considerable wealth, and established an extensive business; while, by their practical operations, in building good serviceable cars and coaches, they have greatly promoted the interests of the traveling and mercantile community. Not by writing down a fanciful theory on paper, as Ross Winans has, but they have promoted the public welfare by a practical exercise of the mechanic art in their branch of manufacture, in a workmanlike and truly efficient manner. Theirs, is an exposition of construction in practice, which the public requires, and which benefits the community. Their earnings are hard earnings; their industry is a practical and laborious industry. Their enterprise is shown by the extent of their works at Troy, and by the fact, that cars, omnibuses and coaches built by them, are all through the north, and different states south, and in foreign countries, contributing to the comfort, safety and dispatch of the public travel. These men are asked to pay only $1,500 for having made and sold, *not* used, the cars which they have, in times past, made and sold; (there is no offer of right to use, but simply to make and sell; the demands for the user are to be princely fortunes, and are to come out of the railroad companies) and $10, or $15, or $30, for the right to make and vend, such as they may hereafter construct; and thus virtually ask them to declare to the public, and the world, and particularly to the railroad companies which have patronized them, that the cars sold as public, were a pirated invention, and render the companies, so far as such a declaration would go, liable for the use of them.

Railroads must use them; they cannot conduct the business of the road without them. They are part of the stock of companies, and their money has been invested in them, as an article of common and public manufacture, as much so, as any other part of the construction and machinery of the road.

If Eaton, Gilbert & Co., were to submit to this exaction, they would, in a measure, render the companies liable, to such an extent as it might have an effect upon other suits brought against them; which would be, not only a forfeiture of that implied faith between them as business men, but if it should be established, the roads might become liable to pay one or two hundred thousand dollars, to be able to use a conveyance which they have purchased from them, without notice or objection from Ross Winans, as an article of public and common manufacture. As honest business men, therefore, they look at this attack upon them as a mere piece of policy, to reach their customers. The toil, trouble and expense of a suit like this, would scarcely be remunerated in the sum of $1,500, which is held out as an inducement to settle. And therefore, considered as a matter of dollars and cents, although more advantageous than to litigate, a settlement on their part is not within the scope of an honorable business transaction, towards those who have purchased their cars, and now use them, for they were built and sold as a common public manufacture.

As to the merits of Ross Winans' patent, they know that the specification is a mere impractical theory on paper, that never has been, and that never can be of any substantial benefit. That the curious style of the specification, as a description of mechanism or mechanical principles, the provisos and conditions, with which it is saddled, provisos fallacious in themselves, and the ambiguous wording of his claim, show that it was not only founded in fallacy, but consciously founded in fraud, as an attempt to pirate a practical principle, known and in use, under the mask of specious distinctions of distance and of construction, both of which are gross errors, and changes departing from the true and known construction and principle of the car—both of which are changes for the worse, and *pernicious* changes in the legal sense of the term. They also know that Ross Winans was not the original inventor of the eight-wheel car, as claimed in this suit by his counsel and himself. That he neither discovered the principle in the abstract, nor the mode of construction by which the principle is embodied to use. And further, these defendants believe, as the proofs here show, that this suit, among others under this patent, is a mere speculation, as I have already said, undertaken by Charles D. Gould; and an attempt to levy black-mail on car builders and railroad companies, under a patent which Winans himself knows is like his class of patents, specious changes for the worse, founded in fraud, and not used by the defendants, and in which he himself has not, and never could conscientiously have had any confidence; so little indeed that while he makes a long affidavit in this case to support his title, he does not swear that the defendants' cars infringe his patent. He does not swear that he invented the Columbus, nor that he made the smoked-drawing before the Columbus was built—though he is aware that they are both matters of controversy, and that they are important facts in this case. He does not swear that the Columbus was built from that smoked-drawing, or that he invented anything in connection with the origination of that car. If these facts are true, he certainly knew it, and why not assert it in his affidavit and under oath, instead of leaving his counsel and partners in interest to come here and say that such are the facts, and to ask this court to believe them? These things are not set up by Mr. Winans; they are set up by those who are experienced in making representations in the trial of patent causes, and I will add, by those who are parties in interest in this suit.

With each and every one of these convictions, these defendants as honorable business men, feel bound to oppose this suit, though the defence cost them as much as the sum for which the plaintiff offers to sell them, that which is of no practical use to them, that is: the specification of this eight-wheel car patent, on which suit is brought. And although they did make a mistake in the form of their first answer, equity will allow them to correct that mistake, because it relates to physical laws and conditions, which neither men nor time can change, and because it relates to a claim, which they are not supposed to be capable of defining, and particularly when the plaintiff's counsel say that it is for one thing, while the claim itself states that it is for another and a different thing. In the Troy case it was for one thing, at Auburn it was for another; here it is for a third, thus carrying two or three faces, one here and another there, just as they may think beneficial to suit the emergencies of the case. Under such circumstances, and, as the error relates to a matter of physical and patent law, the defendants, without the imputation which the counsel on the other side attempted to cast upon them, have a right in equity, to correct a mistake in a single word. It cannot be surprising then, that under such an attack, these defendants should, by their exertions, have discovered a great mass of new matter, explaining to this court the history and true principles of the eight-wheel car, and mode of its construction, and the fraud and fallacy of Winans' specification and pretensions, since the Troy and Schenectady case, and even since the last hearing of this motion at Auburn, in July, 1852. And so arduous are the duties of counsel in patent cases, that they have to go from city to city and collect materials, that have been scattered during years, and which time only will enable them to accumulate, to be prepared in such emergencies.

It may be proper for me here to state, that as some division of the duties arising under this testimony, between my colleague and myself, is necessary, I shall endeavor to confine my remarks to the statement and discussion of the evidence, and leave most of the points of law, and some of the facts, to be discussed by my colleague. Such points of law as I may state will be with a view to indicate the bearing of the evidence, rather than with a view to discuss the abstract principles of law, and the force of their application upon the merits of this case. I shall, however, allude to but few points of law, if any.

The first inquiry therefore, in the mechanics of the case, is: What is the true principle and purpose of the eight-wheel car? And what the necessary construction, to practically develope the principle to use? Upon these points, we have 33 witnesses, out of which number, 28 have been selected, omitting the five who have explained their use of the word *substantially*, in No. 1 of defendants' proofs, and whom the plaintiff alleges have contradicted themselves, although there is no contradiction, but a mere explanation of their use of that word. We have seen from the testimony, that the PRINCIPLE of the eight-wheel car, as built by the defendants, and in general use, is THE SWIVELLING OR TURNING MOTION OF TWO FOUR-WHEEL TRUCKS UNDER ONE BODY, BURTHEN OR LOAD.

The purpose is, to enable the wheels to pursue or conform to the curves and inequalities of the road, and the body, burthen, or load, to be born safely and steadily along the road.

The next branch is:

What construction and conditions are necessary to embody this principle and purpose to use? The first point to be looked to on this subject, is the sum of the conditions that *necessarily* must exist; and construction must be adapted to those necessary conditions.

A necessary condition is, the road itself has and must have curves and inequalities at the joints, on the surface, and at the sides of the rails; also, a lateral play or space between the flanges of the wheels and the rails must exist, to allow of free motion, or prevent binding. This lateral play is generally about one inch IN THE SUM, on the best constructed roads, which would be half an inch on each side. It is sometimes an inch and a quarter, and in an extreme case, would be an inch and a half, which is too much however for good running. About one inch, or as near that as posssible, is the lateral play or space on the best roads.

These are absolute, or constant, conditions as to the road.

As to the car, as a body in motion, there is the absolute law always in force on it when running—"that a body in motion, tends to move in a straight line." That is an absolute condition. Then there are two negative conditions:

1st. The flanges of the wheels must not bite or bind too hard against the sides of the rails.

2d. The wheels must not hop, as it is technically called; or, in other words, must not spring suddenly, by any *cumulative* force, in a transverse direction : that is, transverse to the line of motion.

And, if there be any such cumulative force, or a force variable in its character, and elastic, the wheels will hop, and the pressure will throw them off or over the rails.

Absolutely expressed: the wheels must pursue a course in the line or direction of the rails, without being acted on by transverse cumulative, or transverse elastic forces.

The trucks must be under the ends of the body, without being subject to collision in trains.

The ends of the body must receive the collisions in trains, and not the trucks.

How are these conditions to be carried out, in connection with the principle, so as to embody the principle to use?

What construction of running gear, and what material and mechanism are necessary to accomplish the result, or attain the purpose, by means of this principle of the swivelling motion of the two four-wheel truck, under one body, burthen, or load?

Here is the point where Winans' specification departs totally from the true principle and manner of arrangement and construction, and is worthless, both in its abstract theory, to effect the purpose, and, in the mechanism, to develop the principle.

He, Winans, attempts to establish a new theory as to the proximity of the axles of the wheels, to coincide, as near as possible, with the radius of the curve, in connection with their remoteness, and describes, and especially recommends, certain long springs to connect the axles, as new mechanism, to develop his new theory.

His theory is fallacious, and his construction is pernicious, and trivial, as a piece of mechanism.

We will speak more of this hereafter.

Here Winans' specification, and the cars built by the defendants, part.

One pursues one physical doctrine to develop the principle and effect the purpose, and by certain suitable machinery; and the other, Winans', pursues a different doctrine, by different machinery, to develop the same broad principle, and effect the same purposes; but is erroneous, not only at its base, as abstract matter, but also in the mechanism, and fails, and is pernicious.

What, then, is the necessary construction and machinery to develop the principles and effect the purpose of the eight-wheel car?

We have the principle already stated; we have the absolute and negative conditions already stated; and, we have the effect or purpose, already stated.

The testimony on this subject is very clear on the part of the defence, and by many of the best witnesses that the United States can produce; men of experience—men of the best scientific education and practical training—men from all the branches and departments connected with the construction, the running, and the necessary observation of the workings and of the principle of the eight-wheel car.

Better witnesses the law cannot require; none better can be found in this or any other country. And, we have not failed to collect the experience and the knowledge of all parts of the country upon this subject.

I will refer to some of their testimony. The first I will mention are those who corrected their former affidavits.

Robert Higham, defendants' No. 3, pages 132, 133, 134—fols. 529, and 530 to 534 inclusive; corrected affidavit in defendants' No. 1, pages 41, 42—fol. 271 to 276 inclusive.

Albert Bridges, defendants' No. 3, pages 134, 135, 136—fol. 535 to 537 inclusive; corrected affidavit in defendants' No. 1, pages 36, 37—fols. 235, 236, 237, 238.

Jeremiah Van Rensselaer, defendants' No. 3, page 136—fol. 544 to 548 inclusive; corrected affidavit in defendants' No. 1, page 44—fol. 289 to 294 inclusive.

Isaac Adams, defendants' No. 3, page 138—fol. 550 to 560 inclusive. Troy case; corrected affidavit No. 1, pages 34, 35—fol. 217 to 224 inclusive.

W. Raymond Lee, defendants' No. 3, pages 110, 111, 112, 113—fol. 440 to 451 inclusive; corrected affidavit in defendants' No. 1—fol. 266 to 268 inclusive, page 41.

These are the five gentlemen, out of the thirty-three, who use the word "*substantially*," in their original affidavits, but who, in their amended affidavits, go into an exposition of the principles fully, and give their conclusions on the exposition, and then verify, that the cars which they described have been in use before Winans' alleged invention, as stated in their former testimony.

They have explained the expression "substantially the same," how, under what impression as to the extent of the claim of the patent, and in what sense they used it.

Using the expression "substantially the same," in the sense that two cars have king-bolts with which to have the trucks swivel, is not such an admission or statement as forever to preclude a review of that, as applied to the description and claim in Winans' patent. That is a peculiar description, and a very peculiar claim, the effect of which was not at first apparent.

I will now refer to some of the witnesses, to show what is and what is not, the practical embodiment of the true principle of the eight-wheel car, as built by the defendants, and in general use. J. C. H. Smith, defendants No. 2, page 24, fol. 97; page 25, folio 102, 103.

He says: "I am 43 years of age; reside in Baltimore; am a coach maker by trade, and have been engaged in working at railroad cars since the year 1831, for the Baltimore and Susquehanna Railroad Company."

He is therefore a practical car builder, and belongs to the second or third order of witnesses.

At folio 102, he says: "The proper distance of the wheels apart, to run well on curves and the straight line of road, be-between the flanches, is twenty-three inches for freight cars, and I very often have made them that distance, that is for four-wheel cars; and for eight-wheeled passenger cars they are twenty-one inches apart between the flanches in each truck. This distance is proper to make them run steady, and not press too hard against the out rail of the curves; when the wheels are placed very close together, they do not run steady, but wabble about between the rails, and in striking a curve, the off front wheel bears too hard against the outer rail. The distance of the wheels apart in the trucks of the Tredgold car, is much less than the diameter of the wheels, and the entire description is correct in principle, and sufficient to construct the cars from, that are now in use. I have no interest in this suit."

In order to explain matters as we proceed, I have prepared some diagrams of the cars, such as the defendants make, which I will hand to the court.

I will first state the physical law, "that action and reaction are equal." That is absolute, whenever action takes place reaction must take place somewhere to consume the active force.

Explains the diagrams.

The square of the truck itself, and of the relation of action and re-action, are represented by an equal sided parallelogram, and the guage of the track it will be seen is equal to the square of the truck, or nearly so; it is as near as may be. The leverage therefore, is equal to control the truck on the track. This line, midway between the axles, is the mean radial line of the curve. The line of action is indicated by the arrows. Re-action must take place somewhere, to balance that action—it must be consumed or taken up somewhere. Suppose the line of action to have met with a resistance, and the force exerted to turn the truck to have been a thousand pounds, that amount of force must be consumed or taken up by a re-active force that will balance it, or be equal to one thousand pounds, tending to steady the truck. The mean line of force will necessarily go off midway between the line of action and the line of re-action, intersecting the king-bolt, and balance the truck upon the track, or diminish its tendency to run off the track.

This shows the benefit in having the bearing points of the wheels square, or at an equal distance apart, both across and in the direction of the road. It does not increase the amount of re-active force necessary to keep the truck on the track, over and above the amount of force exerted to turn it off, and hence the actual lateral pressure of the flanges against the rails is not increased or multiplied. It is the increase of the lateral pressure of the flange against the rail, which increases the danger, and a diminution of it diminishes the danger. To have the track square, therefore, balances the force and does not increase either the force or the danger. The longer the truck is made the more safe it becomes, provided the distance between the wheels is not too great for the chord of the segment of the curve between them; they should not bind at the extremities of the chord; they must be free to move forward easily. To diminish the distance between the bearing points of the wheels below a square, that is, to have the actual leverage of the flanges on the same side, taken on both diagonal positions, to be less than the guage of the track, multiplies or increases the reaction or amount of force with which the flange bears against the rail, both because its leverage is diminished and the position of the truck is more oblique to the rail, and hence greatly increases the tendency to run off. Therefore, keeping the wheels square, and bearing on the rails at an equal distance, as near as may be, both across and along the road, balances the leverage in every way around this king-bolt, causes these wheels to remain more nearly in line with

e rails, move with the least possible friction, and be more
eady and safe than when close together.

This figure represents Winans' truck at rest. It is drawn
) the same scale; it shows the mean line of force passing
ehind the king-bolts, which, therefore, exhibits the major
rce exerted to cause the flange to catch and run over the
ail.

The next figure here, shows Winans' truck in motion on a
arve, and exhibits the effects of the excessive reactive or
teral force on the truck; being constructed of springs, they
ust yield to these forces exerted on them; and in addition
) that, the natural tendency of the inner wheel on the
nortest rail is to crowd ahead. This, together with the in-
rease of resistance offered to the front outer wheel, must
vist the truck or these springs more or less out of square, and
acrease its tendency to run off over the rail. This is shown
visted about four inches out of square. The nearness of
ae wheels, and the use of these long springs, when the forces
n a curve are applied, destroy every essential condition that
truck must possess to move with ease and safety on a curve;
ad also, the same derangement is liable on the straight track,
a different degrees, dependant upon the exact circumstances.

These diagrams exhibit the conditions of position and of
ne forces in the defendant's trucks, and in Winans' trucks,
ad display the difference in the mechanical doctrines on
·hich they are applied and operate, and the effects of the
ifferent mechanism employed in them, very clearly.

The next witness I will refer to, is Edward Martin; he is
n important witness as to practical effects. His testimony is
) be found in defendants' No. 2, pages 33, 35—fol. 138 to 145.
He says, in folio 139: "I am at the present time superin-
:ndent of the Schenectady and Troy railroad." At folio 144
e continues: "Of over six hundred cars in use on the line of
ulroad between Troy and Buffalo, Troy and Oswego, and
'roy and Cape Vincent, of which line the road superintend-
d by me forms a part, and which cars pass over our road,
aere is not one having trucks constructed after the manner

of the specfication of Ross Winans. And I would not have a car constructed on that plan, connecting the wheels with springs, nor the wheels in as close proximity as his specification states, as I think it is not practically useful, and not safe. The curves on this line of roads are about as heavy or abrupt as on roads generally, and many of the cars have the wheels in each truck over two feet apart, and all of them have the axles held parallel by substantial wooden or iron frames, and have springs and pedestals attached to the frames, in which pedestals the boxes of the journals move normal to the track." The counsel on the other side, did not seem to understand the meaning of the word "*normal.*" In its general sense, it means natural; and in mechanics, it means perpendicular. For instance: if a railroad were even continued round the earth, it would of course describe a perfect circle; there is no such thing as a perfectly straight or level railroad, and *normal* signifies perpendicular to the curved face, or perpendicular to the face of the rails, as used here. It is an ordinary word, in this sense, in mechanics, and is defined to mean perpendicular. He says further, at folio 145 : " These conditions are necessary, and they could *not* be attained in the plan proposed and described by Winans, in his said specification."

Winans' specification, therefore, will not give the necessary conditions, neither as to the steadiness, nor as to the transverse rigidity; nor as to the perpendicular action; nor as to the relative position of the bearing points on the rails. It is through the bearing points of the wheels that the controlling forces are exerted, and hence their primary importance in relation to each other and to the road.

The next witness is George Beach, defendants' No. 2, pages 36, 37—folios 150, 152, 153, 154, 155. And on page 38, folio 160.

He says: "I am a railroad car builder by trade," and that he is at work for the defendants.

At fol. 151, he says: " I have examined the specification of the Letters-Patent of Ross Winans, of October 1st, 1834, on which this suit is brought, and understand the same. It describes the wheels as close together in each truck as they can be put, without the flanges touching, and the axles to be connected by springs extending from one to the other, bolted to the tops of the journal boxes, and a bolster extending between the springs and securely fastened to the tops of them. The model marked B, and having a paper attached, signed by Stephen Ustick, Wm. A. Aitkin and Oliver Byrne, and on which I also have written my name, is a true representation of the car described by the specification of Ross Winans' patent of the 1st October, 1834. *I never saw a car constructed in the manner described by that specification, in use, on any railroad, and believe that such a car would not be practically useful, as the wheels are too close together.* There is nothing, and no means of applying anything to steady the action of the springs and confine them perpendicular, or normal, to the track; and there is no means for preventing the leaves of the springs from spreading out like a fan."

These springs have leaves, and there is pressure to spread them out transversely. He therefore says:

"There is no means of preventing the leaves of the springs from spreading out like a fan, from the side pressure against the wheels, and nothing in the construction of the truck described, to resist a blow from any obstruction on one side of the road, or a short curve, to prevent the truck from twisting out of shape."

And hence it is, that in action, it assumes this position, by all the forces, and material, and manner in which it is arranged.

Immediately following, he continues:

"It is not safe, and not practical for running on a railroad; it would not stay on the track—it would be unsafe. The wheels of the trucks of an eight-wheel car must not be as close together in each truck as they can be placed without the flanches touching, as they would, if they were thus close, wabble about on the straight track, and injure the road."

There will be a frivolous transverse action; and, as he says, it would "not be as safe as wheels or trucks with wheels as far apart as they can be placed, consistent with the radius of the shortest curves of the road. The wheels in each truck of the eight-wheel cars, are generally placed from twenty to twenty-five inches apart. I have built them myself with the wheels twenty-seven inches apart, and when the wheels are placed thus far apart, the car runs steadily on the straight track, and the outer front wheel in each truck does not bind or bear too hard against the outer rail, and runs easy also on the curves, as well as the straight track.

"I have to-day measured the distance of the wheels, in each truck, apart, of one of the passenger cars run on the Hudson River railroad; they are twenty-one and a half inches apart; the curves of this road are more numerous and of shorter radii than roads generally, to follow the windings of the bank of the Hudson river, and the average speed of running the cars on this road, is at least between thirty-five and forty miles an hour."

At folio 161, he says: "The cars built by Eaton, Gilbert & Co., have a centre cylinder," [this centre cylinder is a transom-plate, commonly so called], "and centre pin or pivot, similar to Tredgold's, and a frame also to support the axles for the wheels to run in; they add the usual pedestal and india-rubber springs, and sometimes steel springs to work in the pedestals, and side bearings of steel or rubber for the body."

We there see that the defendants follow the same construction as in general use. There is the rigid frame and the wheels the proper distance apart, as far as may be consistent with the guage of the track, and the curves and switches on the road, and a centre-plate, in order that the bearing of the car by weight shall keep the truck comparatively steady, and there is also a male and female transom-plate, from 15 to 18 inches in diameter—that is, transome plates provided with flanges, fitting together and forming a large bearing surface and strong centre.

The next witness is L. R. Sargent, as to construction, defendants' No. 2, pages 40, 41—folios 169, 170, 171, 172.

He says: "I am the superintendent of the Rensselaer and Saratoga, and Saratoga and Schenectady roads, and am, and for many years have been, familiar with the principles of construction and operation of the eight-wheel cars built by Eaton, Gilbert & Co., and others, used generally on the railroads.

"I have measured several of the cars now used as follows, and compared the same with the Tredgold car, described and shown in fig. 26, plate 4, pages 179 and 94, &c. in Tredgold's Treaties on Railroads and Carriages, published in London, in 1825.

"The Tredgold car as shown, measures in its proportions as follows:

"Diameter of wheels = 30 equal parts, (considering those parts all through as inches.)

"Distance of wheels apart 25 = equal parts.

"Distance from centre of truck, to end of car = 55 equal parts.

"Distance from centre to centre of axles, = 55 equal parts.

"The Rochester and Syracuse freight car, No. 105, measures in its proportions as follows:

"Diameter of wheels = 30 equal parts.

"Distance of wheels apart = 24 equal parts.

"Distance from centre of truck to end of car = 69 equal parts.

"Distance from centre to centre of axles = 54 equal parts.

"Hudson River Railroad baggage car, measures in its proportions as follows:

"Diameter of wheels = 33 equal parts.

"Distance of wheels apart = 23 equal parts.

"Distance from centre of truck to end of car = 78 equal parts.

"Distance from centre to centre of axles, or bearing points of wheels = 56 equal parts.

"The distance of the wheels apart, of the cars generally in use, varies from 20 to 27 inches, many being more distant apart than the Tredgold car, and of the examples by measurement above given. The distance apart of the bearing points of the wheels of the Tredgold car is one equal part more than the Rochester and Syracuse car, No. 105, and one equal part less than the Hudson river baggage car, they standing 54, 55 and 56 = 55 being the Tredgold, and the centre of the truck is nearer the end of the car in the Tredgold than in either of the others."

Showing that the distance from centre to centre of the wheels, all through, is about the same as the gauge of the track.

The next witness is John Wilkinson, as to general principles of construction, defendants' No. 2, page 43, folio 181, 182.

He says that " he is superintendent of the Syracuse and Utica Railroad Company, as well as president, and is well acquainted with the particulars of construction and action of the eight-wheel cars, as well as of the road on which they run. The diameter of the wheels on the cars, varies from 30 to 33 and to 36 inches." (These are the practical conditions as to the diameter.)

" The distance from centre to centre of the sleepers on which the rails of a railroad lay, were several years ago three feet; and lately heavier loads are carried on railroads, and the distance has been reduced to about two and a half feet from centre to centre of the sleepers on which the rails lay."

There is, of course, an incidental condition, resting upon the weight of the load, as well as the conditions of the road; and it is bad policy to throw too much weight on the same span of the rails, as it has a tendency to crush and bend them.

The next witness is Joseph L. Kite, defendants' No. 4, page 11—folio 41 and 43; page 12—folio 45, 46, 47; page 13—folio 50.

This testimony is taken upon the Quincy car, but the exposition of the principle is embodied in it; hence, I will read it in this connection.

He says, at folio 42: "I have examined the model now before me, marked, on a piece of paper attached thereto, 'Quincy Car, 1829,' and understand it. It has the permanent body to carry the load, with bolsters connected to the centres of two four-wheeled trucks by vertical king bolts.

It also has centre plates and side bearings; and the distance of the bearing points of the wheels in each truck apart, is equal to the guage of the track; and the wheel frames are rigid, to hold the wheel square on the track. This Quincy model car is the same combination and principle as the eight-wheel cars now in general use on the railroads in the United States. The distance of the wheels apart, between their bearing points on the rails, being equal to the guage of the track, is the true principle of the eight-wheel cars as in general use; because the action and reaction between the truck and the rails then deliver their mean line of force to the centre pivot, and balance the truck on the track."

There is the principle, and one of the prominent reasons given, as shown in No. 1 of the diagrams, handed to your Honor.

He continues: "The diameter of the wheels on the Quincy model, in proportion to a five feet guage of track, is twenty inches. Their being smaller in diameter than now generally used, makes the space between the wheels greater than if larger wheels were used."

The diameter of the wheels does not control the principle. The principle is independent of the diameter of the wheels; because the active forces take place between the bearing points on the rail, and they are constant, even though there be a change of diameter.

He continues: "But the difference in diameter, and consequent difference in distance between the wheels, does not effect the essential feature of the distance between the bearing points of the wheels on the rails, as existing in the Quincy car, and in the eight-wheel cars in general use. The diameters of the wheels now generally used are 33 and 36 inches; which, therefore, brings the flanges of the wheel closer together, but does not alter the essential distance between the bearing points on the rails,"

The true principle does not depend upon an incidental distance between the flanges of the wheels. It depends upon the forces exerted on the bearing points of the wheels;

because all the forces combined, and the direction of motion, are due to these points.

He says: " The distance between the wheels is simply incidental to their diameter. The essential distance in the principle of the eight-wheeled car, is the distance between the bearing points of the wheels in each truck on the rails, without regard to the diameter and the incidental distance between the wheels. A change in the diameter of the wheels, and the consequent change of distance between the wheels, is not a change in the principle of the car."

It is not the invention of a new combination, or car, because the principle of the car is the swivelling or turning motion of two trucks under the body or load.

Neither is "a change of the position of the coupling for drawing the car, from the end of the truck to the end of the body, a change in the principle of the car, it is not the invention of a new combination, nor of a new car. Neither would it be the invention of a new mode of drawing a car, for as shown by the printed publications, Tredgold's Treatise and Wood's Treatise, published in 1825, trains of cars were drawn by a coupling from the middle of the end of the body framing, and *not from the axles or running gear*."

The drawing by the body therefore is not the distinguishing feature between a four and an eight-wheel car.

The principle or distinction is not in the change of position of the traction, but it is developed by the two four-wheel trucks, swivelling under the body or load. The witness adds: " It would be no invention of a mode of drawing, to apply the old mode of drawing by the body, to the Quincy car, or in other words, to unscrew the loop from the end of the truck and screw it in the end of the body, would be no invention of a new mode of drawing a car in 1829." Because that mode of drawing had been known and used before.

" A substitution of revolving axles, for the fixed axles in the Quincy car, also would not be the invention of a new combination nor principle."

Because they were both known before. They were both known in four-wheel cars, and whether used in the trucks or sets of four-wheels of an eight-wheel car, or in the sets of four-wheels singly, does not constitute invention, as they were well known in England as well as in this country many years ago, and they are as old as mechanics.

He continues: "Revolving axles and fixed axles were known and used both before and since 1829, and were well known equivalents in 1829. Such change would be only the substitution of known equivalents in the same combination or principle of car, and not an invention, in any sense. To introduce the revolving for the fixed axles, the fixed axles should be removed from the cross-pieces, and the boxes of the revolving axles bolted on to the same cross-pieces of the trucks. Small revolving axles for the purpose, four in number, one of which is marked, on a piece of paper attached, 'Revolving Axles for Quincy Car,' and signed by me, are now before me. They are described on page 13 of Tredgold's Treatise on Railroads and Carriages, published in London in 1825, in the words, 'The wheels of the coal wagons are 2 feet 11 inches in diameter, with 10 spokes, and weigh $2\frac{3}{4}$ cwt., *and their axles are three inches in diameter, and revolve in fixed bushes;*' and on pages 77 and 78 of Wood's Treatise, also published in London in 1825, are these words, 'The axles were made of wrought iron, and fixed firmly into the centre of the wheels, and, consequently, turned upon the bearing with the wheels.' The same, or the revolving axles, are also shown in the plate, No. 51, of the printed publication, 'Reports on Canals, Railways, Roads, and other subjects, made to the Pennsylvania Society for the Promotion of Internal Improvement, by William Strickland, Architect and Engineer, while engaged in the service of the Society,' published at Philadelpaia by H. C. Cary & J. Lea, Chesnut-street, in 1826. The substitution of these revolving axles does not change the principle of the car."

The swivelling motion of the four-wheel trucks, under one body, burthen or load, remains unchanged.

"The principle consists in combining the two four-wheel trucks by means of the permanent body, with bolsters and king-bolts to allow the trucks to swivel to the curves of the road, and this is alike the principle of the Quincy car and of the eight-wheeled cars now in general use on the railroads in the United States for carrying passengers and freight."

The principle, therefore, is in the transverse swivelling motion of these two trucks, under one body, burthen or load, and a mere change of known equivalents as to a mode of drawing, or as to the axles, does not affect the principle, and it still remains in-tact.

The next witness is Wm. Pettit, defendants' No. 4, pages 13, 14—fols. 51, 52, who says: "I am the same person who made an affidavit in this case on the 13th of August, 1852. I have examined the model marked 'Quincy Car, 1829,' and understand it. I have signed the paper attached thereto. It has the two four-wheel trucks with rigid frames, and the wheels in each, apart a proper distance, being, between the bearing points, equal to the guage of the track, and the two trucks coupled with the body and bolsters by means of vertical king-bolts, which allow the trucks to swivel freely to the curves of the road, and being the whole essential principle or difference between the eight-wheel cars now in general use, to turn curves and run steadily, and the four-wheel cars previously in use in this country."

(That is the whole essential difference.)

He continues: "Drawing a car by a coupling from the middle of the end of the body-framing, or using revolving axles with the wheels, is not a new invention. They have both been known and in use, and described in almost all works on railroads prior to 1829, and the substitution or use of them would not be the invention of either a new mode of drawing, nor of adjusting the axles, nor the invention of a new car. I have signed my name also to a paper attached to a small revolving axle, to be substituted for the fixed axle in the cross-piece of the truck frame of the 'Quincy car model,' both this axle and the mode of drawing the car by the coupling from the middle of the end of the body, are shown in the drawing, No. 51, of the printed publication, 'Strickland's Reports,' &c., published in Philadelphia in 1826.

"I am of the opinion that the model marked 'Quincy car, 1829,' is self-evidently the same in its principles of arrangement and action as the eight-wheel cars now in general use on the railroads of the United States."

These portions of the evidence are to establish the principle, as I have stated, and show the necessary construction to embody it to use.

The next witness is Hy. Waterman, defendants' No. 3, pages 64, 65, 66—folios 255 to 263, inclusive.

Mr. Waterman is a practical engineer, and master of machinery on the Hudson River railroad. He ranks among the second class of witnesses. I will read from his testimony, to show what is not a proper construction.

He says: "I have been practically acquainted with railroad machinery since 1837, at which time I took charge of the Hudson and Berkshire railroad. I am the inventor of the dumb, or condensing locomotive for drawing cars in cities, without noise, and which has been for a considerable time used, and is adopted by the Hudson River railroad in the city of New-York. I am fully acquainted with the principles of construction and operation of the eight-wheel cars used on railroads generally. I have examined the specification of Ross Winans' patent, of October 1st, 1834, for an alleged improvement in the construction of cars or carriages intended to run on railroads, and understand the same. The only car actually described, and which is particularly recommended by Ross Winans' specification, is accurately represented by the model B, on the paper of which I have placed my signature. A car constructed as described by Ross Winans, in my judgment, cannot be used with safety for general running on a railroad at an ordinary velocity, such as aimed at by his specification, say at fifteen or twenty miles an hour. His car, as described by his specification, is too dangerous to use on a railroad, both to life and property, for the following reasons: The metal composing the springs that connect the wheels, is liable to constant depreciation of strength by its vibration."

I will here remark, that it is a well known fact that the vibration in metals granulates or weakens them, or weakens them so much that they will break from the vibration.

And he says in substance, that the metal of the Winans' springs, owing to incessant and violent vibration in every direction, " is so liable to break, that it could not be relied upon, and such a break in that truck would certainly do great injury, and probably destroy the life of some of the passengers."

This, therefore, is adverse to the primary purpose of railroads, which is to carry persons and goods SAFELY, and if it be made a means of destruction to goods or the lives of individuals, it is of no practical utility, but detrimental or pernicious in the legal sense.

He goes on to say : " The next reason is, that these spring trucks will not keep square, but loose their rectangular form by the transverse pressure, which will cause the truck to run off of the track. Another reason is that this construction and nature of the truck when attempting to run on a curve of the road, where necessarily the outer wheels have the longest distance to traverse, and bear hardest against the side of the outer rail, allows the axles of the wheels to get out of parallelism with the radius of the curve, by the inner wheels forcing themselves forward of the outer wheels, twisting the truck out of rectangular shape, and necessarily run off the track. It is the worst form of truck that I have ever seen devised.

Another reason why it is dangerous also, is, that the vibration of the springs will allow the axles to get out of parallelism with each other, which adds to its dangerous character before explained. The very close proximity of the wheels required by the Winans' specification, is also erroneous in principle and practice, as it allows a serpentine, unsteady and unsafe motion, both horizontally and vertically, and the wheels follow each other in shocks on the same inequality on the rail, so quickly as to blend the shocks so much together as to become, like the action of a single wheel with a single shock, without giving the front wheel a chance to recover before the next one strikes, and causes the wheels by the concussion and the action of the springs, to hop from the rail and thus run off of the track from this cause, or become dangerous; these inequalities are frequent at the joints of the rails. It is the worst plan for a truck, and all of these causes combined, constitute the very great danger and destructive character. I have in practice proved these principles and actions as stated, to be correct. The proper mode of constructing the trucks of eight-wheeled passenger cars, and that in general use, is to have a rigid rectangular wooden or iron wheel frame, with pedestals securely attached to it, to ensure a normal action of the wheels on the track, and hold the axles exactly parallel to each other, and parallel to the radii of the curves; and the wheels in each truck should be about 55 inches apart, from centre to centre, or between the bearing points of the wheels on the rails; that is, between 54 and 57 inches, with wheels varying from 30 to 33 and 36 inches diameter, which are the sizes in common use. This construction of truck runs with ease and steadiness on both the straight and curved lines of the road, and with perfect safety in the principles and action of the wheels on the rails, at all rates of speed."

The frame being rigid, increase of speed does not affect it; it is fixed in its position and action, and always to be relied on.

The next witness is John Murphy, defendants' No. 4, pages 7 and 8—fols. 27, 28, 29, 30; his testimony shows the principle and construction.

He says: "I am a practical railroad car builder; am a member of the firm of Murphy & Allison, Philadelphia, and am the same person who made an affidavit in this case on the 19th day of August, 1852. I have carefully examined the model now before me, maked on a piece of paper attached thereto, on which I have written my name, 'Quincy car,

1829.' It has two four-wheel trucks connected by a permanent body and centre pivots, with middle bearings and side bearings which allow the trucks to swivel under the body to the curves of the road; the wheels of the model revolve on fixed axles, let into the cross braces of the wheel frame; the distance of the bearing points of the wheels in each truck apart on the rail, is equal to the gauge of the track, or the distance from outside to outside of the flanches. The whole essential principle of the eight-wheeled cars, as used on railroads generally, to carry freight and passengers, known as platform, freight, and passenger cars, is contained in this Quincy model. This essential principle, to turn the curves of the road and run smoothly on it, is the combination of the two four-wheeled trucks, each having the wheels apart between their bearing points on the rails, a distance equal to the gauge of the track, coupled together, by means of a permanent body to sustain the load, with king-bolts, to allow the trucks to swivel under the body to suit the curves and inequalities of the road. The trucks and body have the bolsters, side bearings and the rigid rectangular wheel frames, to keep the wheels in their position on the track, being the same practical conditions as now existing in the eight-wheeled cars now in general use on the railroads.

"In the early days of the railroads, the cars were made with wheels both to revolve on fixed axles, and also with the axles to revolve with the wheels. I have fitted them up both ways myself."

This is the testimony of a practical car builder, who, although he is acquainted with the plaintiff, and has been upon business and friendly terms with him, still does not hesitate to say that this is the true principle of the eight-wheel car.

The next witness is Christian E. Detmoold, defendants' No. 3, page 56—folio 224; pages 59, 60—folios 236, 237, 238.

He says: "I am 42 years of age, and am a civil engineer by profession; at present I am the superintending engineer of the New-York Crystal Palace, about being erected, and am connected with several other engineering enterprises. In the summer of the year 1828, I was engaged in the service of the South Carolina Railroad Company, and continued in their service until 1833."

At folio 236 he continues, after speaking of the existence of the eight-wheel steam carriages:

" These steam carriages contain the same substantial combination of two four-wheel trucks, and long carriage body, by means of centre pivots and bolsters, as the eight-wheel carriages or cars for transporting freight and passengers on the roads now in use, on the railroads generally. These carriages, like the South Carolina and the original drawing G, also have rigid wheel frames, and pedestals, and springs, to properly control and regulate the action of the wheels in running on the road."

The principle was not a new thing then, for it was embodied practically and fully in those steam carriages.

" In every essential respect and particular, and in the substantial character of the combination of the body with the trucks, the eight-wheel cars now used generally, and the eight-wheel steam carriage, of which Mr. Allen and myself, under his directions, made the drawings in the winter of 1830 and '31, are the same. Putting a box or seats on the steam carriage for passengers, and making it a steam passenger coach, or removing the boiler and setting a box on the upper bolsters for passengers to sit in, does not alter the substantial character of the combination, and does not produce a new invention. It would still be the same substantial combination, of a long carriage body connected to two four-wheel trucks by bolsters and centre pivots with side bearings, to conform to the curves and follow the track with ease and safety, and developes, in both cases, the same principles of operation and the same advantages both to the road and the carriage body. The distance of the wheels apart in each truck of the original drawing, and of the steam carriages, was twelve inches."

The bearing points of the wheels, therefore, are equal to

the guage of the track. This testimony shows both the principle and the construction or mechanism in practice.

The next witness is Wm. J. McAlpine, defendants' No. 3, pages 92, 93—fols. 368, 369, 370, 371; page 94—fol. 374; page 95—fols. 379, 380.

Mr. McAlpine is one of the first class of witnesses, and he is not only among the first class as regards his education and profession, but also by practical experience.

He says: " I am 42 years of age, and reside in Albany; am civil engineer by profession, and have been engaged in such occupation through a period of about 27 years. I have been engaged on the Carbondale railroad ; the Mohawk and Hudson railroad ; the Schenectady and Saratoga railroad; on the St. Lawrence improvement; the Chenango canal; the enlargement of the Erie canal; was engineer of the U. S. Dry Dock, at the Brooklyn Navy Yard ; was engineer under Chancellor Walworth, in the Wheeling bridge case, referred to him by the Supreme Court of the United States, and afterwards was appointed by the court to examine and report engineering facts in the case; and I was engineer of the Albany, and also of the Chicago Water Works, and at the present time am the state engineer and surveyor of the State of New-York. I am acquainted with the principles of construction and operation of railroads, and of the eight-wheel cars run upon them. I have examined the specification of Ross Winans' patent of October 1, 1834, for an alleged improvement in cars or carriages, intended to run on railroads, and understand the same. The model B, on the paper to which I have written my name, is a correct representation of the car described in detail, and recommended by Mr. Winans. I do not deem it an improvement upon the eight-wheel cars or carriages, previously known for the same purpose. It is, in my opinion, comparatively unsafe. The wheels in each truck are too close together. The axes of the wheels will not remain parallel to each other, and the wheels in each bearing carriage will not keep square in running on a curve, which would tend to throw the car off the track, and are difficulties in its principles and details of construction, and consequent action, so serious as to render it unsuitable for general use, and unable to compete with the eight-wheel cars or carriages having the rigid rectangular wheel frames and centre pivots, previously known and now extensively used on the railroads generally."

He then goes on to describe Chapman and Tredgold, which will not read in this connection.

At folio 374, he continues: " In that description and drawing of Chapman, we have the essential principle of construction and operation, both in detail and organization of the eight-wheel car now used. There is the rigid rectangular wheel frame of four wheels, with the cross-piece, centre pivot and side bearings, and the bearing points of the wheels, whether larger or smaller wheels, may be used, given a distance apart about the same as the breadth of the track, and the carriage body sustained by the two bearing carriages, laying the weight equally, being the same essential principles of construction as used at present in the eight-wheel car."

On page 95, folio 379, he says: "And also the wood cars described by Conduce Gatch, Jacob Rupp, and Leonard Forrest, in their affidavits in this case, are the same principle of construction and organization, and in all essential details the same as the eight-wheel cars now in common use. Putting a box with seats on the bodies of the wood cars, instead of the wood, would not form a substantially new combination, nor involve any invention. No new principle would be developed, and with reference to the South Carolina steam carriages, removing the boilers or body, and securing a box with seats on the upper bolsters in place of them, would not form substantially a new combination. It would simply be the analagous use of known things, for a known purpose, forming the same substantial combination, and not an invention."

So that it is clear that the principle does not consist in putting on boxes in place of boilers, and that when the swiveling motion takes place, the principle is developed without regard to the character of the body, or whether a box or platform, finished in one style or another, rests on the bolsters.

The next witness is Chas. B. Stuart, defendants' No. 3, page 89—folios 353, 354, 355, 356; page 90—folios 357 to 359; page 91—folios 362 to 364; page 92—folios 365, 366.

Mr. Stuart's whole testimony shows that he speaks from practical experience.

He says: "I am 38 years of age, and reside in the city of New-York, and am an engineer by profession. I am, and have for the past twenty-two months, been engineer in chief of the United States Navy. Have been state engineer and surveyor for the State of New-York. Have been. chief engineer of the United States dry dock at the Brooklyn navy yard; and have been connected with railroads for twenty-one years; and seventeen years of that time was constantly engaged on them. I was assistant engineer on the Saratoga and Schenectady railroad; also the Saratoga and Whitehall railroad, and the Utica and Schenectady railroad. I was resident engineer on the Syracuse and Utica railroad, and was chief engineer of the Oswego and Syracuse railroad; chief engineer of the Tonawanda railroad; chief engineer of the Susquehanna and central divisions of the New-York and Erie railroad; chief engineer of the Rochester and Niagara Falls railroad; chief engineer of the Great Western railroads from Niagara Falls to Detroit; chief engineer of the Chemung railroad; and have been connected with other roads. I have had very extensive experience in the construction and operation of railroads; and I understand the principles of construction and operation of the eight-wheel cars run on them. I have carefully examined a certified copy of the specification of Ross Winans' patent, of October 1st, 1834, for an alleged improvement in the construction of cars, intended to run on railroads, and the model B, on the paper of which I sign my name. The model conforms to the specification, and is such as a mechanic would make from the description. I do not think the car, described by Mr. Winans' specification, safe or fit for practical use on railroads; and it s entirely different in the details of its construction, and in the alleged important principles, set out in the specification from the eight-wheel cars, now in general use on railroads; and no mechanic would or could make the eight-wheeled cars, now in general use, from Winans' specification. I think the specification purely theoretical, and unfit for practical purposes at an ordinary or usual rates of speed. I have carefully examined the description of the eight-wheeled carriage, contained in Tredgold's Practical Treatise on Railroads and Carriages, printed and published in London in 1825, and the model A, and understand them. The descriptions on pages 94 and 179, and the plate 26, fig. iv, is a complete description of the eight-wheeled car, with two four-wheel frames, and vertical axis, to allow them to turn between the rails to the curves, and all the other inequalities of the rails, conform to the changes of level, and divide the pressure equally among the wheels, as described and required by Tredgold; and the model A, on the paper of which I have signed my name, is a correct representation of it. A mechanic could make the eight-wheeled cars now in use, in all their essential particulars, from Tredgold's description, without any other aid whatever. It contains all the essential principles for practical operation, and is far more correct in its details for practical purposes than the Winans' specification. Putting the longer body A 2, on the trucks of model A, would not alter the principle. All the essential principles of the car would remain the same."

Then, on page 91, folio 362, he adds: "I have examined the specification and drawing of a patent, granted to W. Chapman, civil engineer, and E. Walton Chapman, of England, dated December 30th, 1812, as printed and published in the Repertory of Arts, Manufactures and Agriculture, vol. 24th, 2d series, February, 1814, published by J. Wyatt, London, England, and find there described a truck of four wheels, with rigid frame, and the bearing points of the wheels a proper distance apart on the rails, about equal to the breadth of the track, to be placed under the ends of railway carriages; one of them with a single pair of wheels, for a six-wheel carriage, similar to the six-wheel American locomotives, and two of these trucks for an eight-wheeled carriage if desired, similar to the eight-wheeled carriages now used, for the purpose of allowing the carriage to move freely around curves, or pass the angles of a railway, allowing the frames to turn on a pivot and distribute the weight equally on the track, as is common in the cars now in use in the United States generally. An ordinary mechanic would be able to make the eight-wheeled car from that description. The details of construction and arrangement there shown, are the true and practical principles of the eight-wheeled car. The description is much more practical and better in principle than the Winans' specification."

Then on page 92, folio 365, he continues: "Also, the wood cars, described by Conduce Gatch, Jacob Rupp and Leonard Forrest, in their affidavits in this case, have the two four-wheel trucks, rigid wheel frames, vertical pivots, and long permanent body, connecting them by bolsters and the pivots, the same identically in principle of construction and operation as the eight-wheel freight and passenger cars now in use on railroads generally. The main details of construction, and the principle of the combination of the two four-wheel frames with the long body, by vertical axis and cross pieces, to allow the wheel frames to turn between the rails, is a very old thing, as now used, forming the eight-wheeled cars on the railroads generally. And Winans' spring trucks, coupling the wheels closely and remotely, as he describes, is no improvement."

The next witness is Vincent Blackburn, defendants' No. 3, pages 96, 97—folios 382, 383, 384, 385; page 98—folios 391, 392.

Among other things, he says that the change of the diameter of the wheels does not alter the principle. In other words, the principle is not involved, or does not lay in the diameter of the wheels.

He says: "I am a machinist, and have charge of the machinery department of the Utica and Schenectady Railroad, and have been engaged for twenty-one years in the construction of railroad machinery. I am well acquainted with the principles of construction and operation of the eight-wheeled cars, used on railroads generally. There are certain details and principles of construction in the eight-wheeled cars, that are necessary to give a secure and practical operation to them on the road, and these details of construction and combination, composing the car, are very old. For example, the breadth of the track of the line of railroad, between Albany and Buffalo, is four feet eight and a half inches, (fifty-six and a half inches) and the distance between the bearing points of the wheels in each truck or wheel frame on the rail, is about the same, or equal to the breadth of the track. The distance between the bearing points of the wheels on the rails is the essential distance, and not the distance between the wheels, or the faces of the wheels, as this latter distance will vary with the variance of the diameter of the wheels used; the bearing points remaining the same distance apart, or the wheels the same distance from centre to centre. The distance between the bearing points or centres of the wheels of the trucks of the eight-wheeled cars, are and should be about the same as the breadth of the track used, forming a square or figure of equal distance across the track, and parallel to the track, between the bearing points. The wheels should not be placed as close together as they can be without touching each other, as stated by Ross Winans, in his specification of Oct. 1, 1834, which I have read. In the construction of the wheel frames or trucks, as they are commonly called in the United States, used on the eight-wheeled cars generally, the axis of the wheels are held parallel to each other, by rigid rectangular wheel frames, having centre pivots, the same in principle as the wheel frame or truck shown and described in fig. viii, plate v, in Chapmans specification, printed and published in the Repertory of Arts, volume 24th, 2d series, 1814, in which also the bearing points of the wheels are a distance apart, about the same as the breadth of the track, and two of these trucks are described for an eight-wheel carriage, and one with a pair of wheels for a six-wheel carraige."

Then on page 98—folios 391, 392, he continues : " I have examined the drawings H and G, signed by Horatio Allen, and others, of the South Carolina engines or steam carraiges, and they contain the same essential organization and combination to turn curves, that now exists in the eight wheeled carriages in general use, many of which are constructed by Eaton, Gilbert & Co. of Troy. The essential principle of the combination of two four-wheel rigid wheel frames, with pedestals and springs, centre pivots and side bearings supporting the long carriage body, by means of bolsters, is all there, the same as used in the eight-wheeled freight and passenger cars in general use. I have examined the specification of Ross Winans' patent fully, and he particularly describes and recommends a car or carriage there, similar, or the same as the model B, signed by me. It is no improvement whatever, over either Chapman's, or Tredgold's, or Allen's carriages. This mode of construction would not suit for the Utica and Schenectady, and the other roads generally. It is not used. It would be too unsafe. I would not recommend it on any account."

Here is the testimony of a superintendent of a road, having power to put machinery on the road ; but who says that he would not use such a thing, as it is incompatible with the construction and principle of an eight-wheel car.

The next witness is Walter McQueen, defendants' No. 3, pages 99, 100—folios 395, 396, 397, 398, 400, 401 ; pages 102, 103—folios 408, 409, 410, 411 : " I am a machinist, and have charge of the Schenectady locomotive works. I have been engaged in the construction of railroad machinery for about thirteen years. I am the designer of the locomotives on the Hudson River railroad, that run the express trains at from forty-five to sixty miles an hour. I had charge of the machinery of that road prior to taking charge of the Schenectady works. I am well acquainted with the principles of construction and operation of the locomotives, and of the eight-wheeled cars for conveying freight and passengers used on the railroads generally in the United States, and the details necessary to the safe and proper practical operation. It is necessary that four-wheel trucks, to run on a railroad with safety and ease of motion, should have rigid rectangular wheel frames, holding the axis of the wheels parallel, and the bearing points of the wheels square, so that the truck shall not be twisted out of shape, and the engine or car run off of the track and overturn. To secure the proper action in the eight-wheeled cars constructed by Eaton, Gilbert & Co., and such as are used generally on the railroads, the wheels are connected with each other, and the body of the car by means of rigid rectangular wheel frames, that hold the bearing points of the wheels in each frame, a distance apart about equal to the breadth of the track, and the axis of the wheels parallel to each other, and keep the wheels square in running on curves; or at other inequalities such as depressions or elevations at the joints of the rails, while the truck turns between the rails from the resistance offered by the inequalities, at which time in all cases, the axis of the wheels are not in the same plane, whether on the straight or curved part of the road. This is the essential arrangement and principle of the eight-wheeled car. The distance of the trucks apart, depends upon the length of the car, and will vary with a variation of the length of the body, whether shorter or longer, without changing the essential principle. This essential principle of the eight-wheeled carriage, whether it has horses, or men, or steam in it, to move or draw it, together with freight or passsengers, or is used for freight or passengers alone, was given and is to be found, described and shown in Chapmans' specification, contained in the 24th volume, 2d series of the Repertory of Arts, printed and published in London, in 1814. That specification and drawing, on pages 130 and 139, and fig. viii, plate v, distinctly describes and shows both the six-wheel and eight-wheeled carriage, in all the essential detail and organization or combination, as at the present time running on the Utica and

Schenectady railroad, and also on the Albany and Schenectady railroad, to turn the curves, divide the pressure on the wheels, and run safely, smoothly and in a direct course on the track. I constructed some of the six-wheeled carriages for the company to carry wood and water; and the eight-wheeled carriages are the same as constructed by Eaton, Gilbert & Co. of Troy, and such as are used on railroads generally, to convey freight of all kinds, and passengers. The truck or four-wheel frame, in fig. viii, and as described, has the rigid rectangular wheel frame, holding the axis of the wheels parallel, and the wheels square on the track, and the distance between the bearing points of the wheels is about, or exactly the same as the breadth of the track; has also the centre pivot and side bearings; and two of these, both alike, and in the same way support the carriage body, under which they turn on the pivots to the curves and other inequalities, the same in arrangement and principle as the eight-wheeled carriages now used on railways."

Then at folio 408, he adds: " The same arrangement and principle of construction and operation, as contained in the eight-wheeled cars, is also contained in the steam carriages, delineated by the drawings G and H, signed by Horatio Allen and others, and by myself. They have the long body secured to, and borne by bolsters or cross-pieces, connected by centre pivots to two rigid rectangular wheel frames, having each four wheels, with the axis parallel, and pedestals, and springs, and side bearings; and the bearing points of the wheels on the rails are square, or equal to the breadth of the track, as shown by measurement of the general plans in both drawings. The arrangement, combination and principle, is identically the same as the eight-wheeled cars or carriages now in use, in all the essential details and organization; and that steam carriage, or a box placed on the same bolsters, to contain freight or passengers, would move past the curves and angles of the railway with the same facility and safety, and pursue a smooth, even and direct course on the road, the same as the eight-wheeled cars now generally in use.

"I have examined the specification of Ross Winans' patent, of October 1, 1834, and understand it. It does not set forth any fixed principle, nor any specific mode of construction or organization, from which a car can be constructed to run with safety or practical utility on a railroad. It describes the wheels as close together as they can be placed, without stating what diameter of wheels are to be used, and without giving any idea of the proper distance of the bearing points of the wheels apart. It does not describe any mode, nor intimate any means or necessity of keeping the axles parallel, nor the wheels square on the track. It describes a spring coupling the wheels, which is wholly impracticable, will not keep the axles parallel, nor the wheels square. I do not consider it safe to run upon a railroad. No such car, to my knowledge, is in use upon any of the railroads of the United States. The model B, to which I sign my name, is a correct representation of the car 'intended to run on railroads,' described by the specification, and patented as such by Mr. Winans."

We consider Mr. McQueen one of the ablest and most experienced practical engineers. His affidavit contains a large amount of practical information on the points of construction and principle, hence I have referred extensively to it.

The next witness is George S. Griggs, defendants' No. 3, page 104—fols. 414, 415, 416, 417, 418, 419.

He says: "I am the master mechanic of the Boston and Providence railroad, and have the control of all the construction department of machinery, consisting of cars and engines belonging to the road. I have been master mechanic for eighteen years on the same road. I have had the charge of construction of all the engines on the road, and we have built many of our eight-wheel cars. The first eight-wheel cars built by us in 1838, are now used on the road. The width of the track is 4 feet $8\frac{1}{2}$ inches between the rails; and the distance between the bearing points of the wheels, in each truck on the rails, is about 5 feet; being more than the breadth of the track. The wheels are 33 inches in diameter, and the distance, therefore, between the wheels, is about 27 inches. Their trucks are made with rigid rectangular wheel frames, as stiff as we could make them, and braced to strengthen them and keep the wheels square and the axles parallel, and thus enable the cars to run well on curves, and also the straight track. All of the bearing carriages of the eight-wheel cars that have ever been used on the Boston and Providence road, have rigid wheel frames, and the bearing points of the wheels apart differ but a few inches. As a general rule, the distance of the bearing points of the wheels in each wheel frame apart, are about the same as the breadth of the track. I measured two of the passenger cars at the Roxbury station this morning; they have been running on the road for about six years, and the turn-outs, and switches, and curves of the road are as sharp, as on roads generally. One of these cars measured 4 feet 9 inches between the centres of the axles or the bearing points of the wheels on the rails in each wheel frame; and the other measured 4 feet 10 inches between the centres of the wheels on the bearing points of the wheels on the rails in each wheel frame, being, in both cases, more than the breadth of the track. On the first eight-wheel cars built in 1838, the wheels were 36 inches in diameter, and so used for about seven or eight years, when 33 inch wheels were placed on the same cars, bearing the centres of the wheels or bearing points on the rails the same distance apart as before; 33 inch wheels are now used on all the cars. The distance of the centres of the wheels, or the bearing points of the wheels on the rails apart, in each wheel frame, is the essential distance in the eight-wheel cars for turning curves and running on the straight track. The distance of the faces or flanges of the wheels apart is not the essential distance, as it will vary with a variation of the diameter, and the wheels should not be placed as close as they can be without the flanches touching, as a general rule. The distance of the centres of the wheels apart, or their bearing points on the rails, should always be about equal to the breadth of the track, which gives a good leverage to keep the truck on the track and make it run steady on both curves and the straight line of road. These trucks are coupled apart any distance required, according to the length of the car, by the centre pivot and the body of the car."

There he gives, not only an exposition of the principle, but he shows the practice in regard to the principle, and that the principle remains the same, although there may be a change in the diameter of the wheels, and consequent changes in the distances between the wheels.

The next witness is Albert S. Adams, defendants' No. 3, pages 114, 115—folios 454, 455, 456, 457.

Mr. Adams is a witness of the second class, and is the master machinist of the Boston and Worcester railroad, a road which does the heaviest business in the New England States, and where the greatest amount of work is performed, and therefore the probability is, that these things are more fully tested there than elsewhere.

He says: "I am the master machinist of the Boston and Worcester railroad, and have been connected with railroads as machinist for fourteen years, and understand the principles of construction and action of the eight-wheeled cars now in general use on railroads. The eight-wheel cars in general use, have and necessarily must have to be practically safe and useful, rigid rectangular wheel frames, with centre pivots or bearings to support the body, and allow the frames or trucks to turn to the curves, and turn from the resistance of inequalities of level on the straight, and also curved line of road. The rigid frames hold the axis of the wheels parallel, and the wheels square, or their bearing points rectangular on the track; and these trucks are secured a sufficient distance from the ends of the car, so as not to be damaged by collision of cars running together, and the truck sufficiently far apart to distribute the weight over the different rails of the road as much as the length of the body that may be used or desired, will allow. The best practical arrangement of the wheels apart in each truck, is to have their bearing points on the rails a distance apart about equal to the breadth of the track, without regard to their diameter.

"I have read the specification of Ross Winans' patent, of October 1st, 1834, for a 'car or carriage intended to run on railroads.' His theory is to place the wheels in each truck as close together on each side as they can be without their flanges touching, 'to be considered as acting like a single wheel.' The fact is, that a single pair of wheels with a pivot would turn around and lay lengthways between the rails, and this same tendency of his wheels close together, tends to render them unsafe, or too unsteady; and his means or manner of constructing these trucks by springs, coupling the axles of the wheels, makes them so unsafe that they would be unfit for practical use by the public on the railroads. The model B., signed by me, is a correct representation of the car described by the specification of Mr. Winans' patent. I am of the opinion that his car would not be safe or practical for use on a railroad, to carry either freight or passengers. It would be extremely dangerous to the lives and limbs of passengers."

The next witness is James H. Anderson, defendants' No. 3, pages 117, 118—folios 467, 468, 469, 470, 471; pages 119, 120—folios 476, 477.

He states that he is "the master machinist of the Providence and Stonington railroad, and has charge of the construction of the cars and engines of the road. I have been engaged as a mechanic, on cars and machinery, for about ten years previous to becoming master machinist. In all, I have been engaged about twelve years. I understand the principles of construction and action of the eight-wheeled cars in use on railroads generally, and have constructed and understand the action of the trucks.

"The breadth of track of the Providence and Stonington road, is four feet eight and a half inches. This is the usual guage of railroads, and the distance of the bearing points of the wheels apart in each truck of the eight-wheeled cars on the rails, is about equal to the breadth of the track. In some of them the distance is five feet, being a little more than the breadth or guage of the track. The trucks are rigid wheel frames, holding the axles of the wheels parallel, and the wheels square on the track.

"It is necessary, to the safety and good running of an eight wheel car, that the trucks be constructed of rigid rectangular wheel frames, holding the axles parallel, and the wheels square on the track, then the friction in running on curves is the least, and the car moves smoothly and safely.

"I have examined the specification of Ross Winans' patent, of Oct. 1, 1834, and the model B, signed by me, is the same in construction as described by the specification, as the thing patented. The specification of the patent, requires the wheels to be too close together for safety and smooth running of the car, and would cause great friction on both curves and the straight track. Also the connection of the wheels by springs, would be insufficient or impracticable for railroad purposes; it would not hold the axles parallel, nor the wheels square, and would not be safe.

"The manner of construction and of connecting and arranging the wheels of the trucks of Winans' car, is essentially different from the cars now in use by railroads generally; and its action, resulting from the connection and arrangement, would make it unsafe, unsteady, cause too much friction, and unfit for railroad purposes."

This clearly shows what is and what is not the true construction to develop the principle to use, and on this head I will refer to the testimony of a few more of our witnesses, in order that all the practical information on the subject may be fully brought forward.

The next witness, whose testimony I will read, is Wm. C. Young, defendants' No. 3, pages 149, 150, 151, 152—fols. 597, 598, 599 to 600; and fols. 604, 605, 606.

He says: "I am 52 years of age, and reside in the city of New-York. Am an engineer by profession; was educated, and graduated at the United States' Military Academy, at West Point, and at present am the President pro tem. of the board of directors of the Panama Railroad Company. In

1831, and until 1833, I was a resident engineer of the Saratoga and Schenectady railroad ; in 1833, and until 1849, I was the chief engineer of the Utica and Schedectady railroad, and in 1849, until 1851, I was the chief engineer of the Hudson River railroad. I have had practical experience in the construction and use of the eight-wheel cars used on railroads generally, and understand their principle of construction, and their action in running on the road.

" The model B (signed by me), is a true representation of the car described by the specification of Ross Winans' patent, of October 1st, 1834. Placing the wheels as close together in each truck, as he describes as a material part of his invention, is objectionable. An eight-wheel car runs more steadily and with less friction with the wheels in each truck a considerable distance apart, than when close together. I have placed them six feet apart from centre to centre, or between the bearing points, and they run more steady, and with less friction on curves, than when placed close together; a distance apart between the bearing points, equal to the guage of the track, which is generally four feet eight and a half inches, is close enough for steadiness, safety, and avoidance of friction in running on curves, and through switches.

" The mode of connecting the wheels with springs, described by Winans, is too frail and yielding to be safe and practical for railroad purposes. It is unsafe, and not useful.

" Tredgold's Treatise on Railroads and Carriages, published in London in 1825, clearly describes the construction and action of an eight-wheeled car on a railroad, similar to model A, signed by me, on pages 94 and 179, and fig. 26, plate iv. It is a true representation of Tredgold's description, which is within the capacity of any ordinary railroad machinist, in my opinion, and is the same in principle as the eight-wheeled car now in use, and the same in the essential detail to carry out the principle. It performs all the functions required by Tredgold, on the pages and figure above referred to.

" The model C signed by me, is not described by Tredgold, and does not perform the functions he requires, and it would not suit for a railroad car.

"The eight-wheel car, in its essential construction and organization, was described before Tredgold's publication, in the specification of Wm. and E. W. Chapman, in the Repertory of Arts, volume 24, second series, published in London in 1814. The model K (signed by me), represents it, and also the six-wheel car, being changeable from one to the other.

"Both of them are in use on several of the railroads between Albany and Buffalo, and the eight-wheel is in general use. Both the Tredgold and the Chapman car, have the rigid rectangular wheel frame, the axis of the wheels parallel, and the bearing points of the wheels about equal in distance apart, to the guage of the track, and the swivel on separate axis or pivots under one body, to suit the curves and irregularities of the rails; or as Tredgold more comprehensively expresses it, 'turn when (or at the time) from any inequality (or level) the axis of the wheels are not in the same plane.' This occurs both on the straight and curved track, and these principles of action are those of the eight-wheel cars in general use.

"Placing the same trucks farther apart, with the same or a longer car body, or placing the truck wheels closer together and the trucks farther apart, is no invention; it is simply a change of proportion.

"The whole principle and arrangement of the eight-wheel car now used, is practically organized, and clearly stated, and shown in both the description of Chapman and of Tredgold.

"The same essential principle of construction, arrangement and action of the eight-wheel car, is contained in the steam carriage, delineated by the drawings G and H, signed by Horatio Allen and others, and by myself. The body resting on the upper bolsters, being a steam boiler, does not change the substantial combination or arrangement. Putting seats on it, or a box on the bolsters in place of it, to carry freight or passengers, would not change the substantial character of the carriage, to turn curves, and pursue a steady and safe course on the road. The invention would be the same. It has the bolsters, centre pivots, rigid rectangular wheel frames, parallel axis, and the bearing points of the wheels equal apart to the guage of the tracks, and supports the body by a truck under each of its ends, being the entire essential principle of the eight-wheel cars in use."

The next witness is Godfrey B. King, defendants' No. 4, pages 79, 80, 81, 82—folios 298, 299, 300 to 310 inclusive. In his affidavit this whole subject of construction and principle is so fully treated, that I will refer largely to it.

He says: "I am the master machinist of the Boston and Lowell railroad, and have been on the road ever since it was started, being for a period of about seventeen years; and I designed and built the engine that took the premium for draught, at the trial of locomotives in New-England, which took place at Lowell about one year ago.

"I am acquainted with the construction and action of all the eight-wheeled cars that have been used on the Boston and Lowell railroad. We frequently construct our eight-wheeled cars. They are on the same plan of construction and operation as those generally in use on railroads. I have examined the specification of Ross Winans' patent, of October 1st, 1834, for a car or carriage intended to run on railroads, and understand it. The Boston and Lowell railroad never used a car constructed in the manner, or upon the plan described in Ross Winans' specification. The cars that have been used by the Boston and Lowell road, have the bearing points of the wheels on the rails about the same distance apart as that of the rails of the road, forming a rectangle or square of equal leverage about the centre pivot, and the wheel frame of the trucks, is the usual rigid rectangular wheel frame, holding the axles parallel and the wheels square on the track, both on the curves and on the straight lines of the road, the same in principle of construction and action, and also the same in arrangement and organization with the car body, as the eight wheeled car described in the 24th volume of the Repertory of Arts, second series, in Chapman's specification, published in 1814; and the same also, as contained in Tredgold's Treatise on Railroads and carriages, published in London in 1825. The model K, signed by Wm. B. Aitken, and others, is a correct representation of both the six and eight-wheeled carriages described by Chapman, in the 24th volume of the Repertory of Arts, as it illustrates both the six and the eight-wheeled carriages, being changeable for that purpose. The six-wheeled carriages to convey wood and water, and also the eight-wheeled carriages there described, are in use in this country.

"The carriage may be moved by any kind of power, in various ways, of different degrees of utility; and the long body or single body, sustained by the two four-wheel trucks, properly constructed and acting, forming the essential principle of the car, still remains the same, though the mode of drawing or moving the car may be changed. Its construction and arrangement, as described by Chapman, like the eight-wheeled cars now in use, is adapted to being drawn by the middle of the end of the body, because he shows the necessary wheel frames, and the bearing points of the wheels in the trucks on the rails, a distance apart equal to the breadth of the track, and, therefore, the trucks have the bearing points of the wheels, in each of them, sufficiently far apart and firmly held, to enable the flanches to keep the trucks steady, when left free on their centres, as they are when the car is drawn by the body.

"The model A (signed by Oliver Byrne and others), is an accurate representation of the eight-wheeled car described in the book, Tredgold's Treatise on Railroads and Carriages, published in London in 1825, on pages 94 and 179, plate iv, fig. 26. It has the same essential construction of the rigid wheel frame; the wheels, as shown by their diameter and their proportionate distance apart in each frame, having their bearing points a distance apart about equal to the breadth of the track; the vertical axis, to enable them to turn at the time, or when the axis of the wheels is thrown out of the same plane by the change of level or inequality, which is incessant on both the straight track, and particularly on curves, and which at the joints causes the constant clicking noise to be heard as the cars run. These rigid wheel frames, of the Tredgold car, are also shown, arranged and organized with the body, as near as they can be, to the extreme ends of the body, and still have those ends protect the trucks from destruction or injury in the collision of cars running together; and therefore, of course, as far apart as they can be placed, consistent with the length of the body, to distribute the stress of the wheels upon the different rails of the road. All the essential detail of construction, and arrangement, and connection of the wheels with each other, and with the body, and also the action, safety and steadiness of motion that exist with the eight-wheeled cars, are contained in the description and drawing in Tredgold's book. That is, all the essential features are there sufficiently set forth to enable an ordinary railroad machinist or car builder, to construct and use from it the eight-wheeled car now in use, with any particular length of body that may be desired. The construction of the Tredgold car, as shown or described, is particularly well adapted to being drawn by the body, in the manner now practised, as the trucks are capable of steadying and controlling themselves by their flanges being a proper distance apart, and drawing by the body with a coupling at the middle of its end, is the method shown in the same book, plate 1, as the established method of drawing cars in trains, and no directions being given to change it, a mechanic would use it the same.

"The steam carriage delineated by the drawings G and H, signed by Horatio Allen, C. E. Detmoold and others, contains the essential combination and construction of the eight-wheel carriages used generally to carry wood and water, and freight, and passengers, separately or together. It has all the necessary construction of wheel frame or truck, with the bearing points of the wheels held the proper distance apart, connected by the proper kind of frame, and the two wheel frames connected, one under and to each end of the body, by the separate vertical pivots and body bolsters—substituting a box to carry freight or passengers on the upper bolsters, in place of the iron body or boilers, would not involve invention, nor form substantially a new arrangement, or organization. The mere fact that passengers could stand or sit more conveniently in one than about the other, would not form or constitute the combination of a new one, and no new principle would be developed.

" The car or carriage described by Mr. Winans' specification, is accurately represented by the model B (signed by Oliver Byrne and others.) It is such in all respects as I should make from the description. He, Winans, does not attempt to describe any particular mode of drawing it, or moving it on the road; but silently leaves the selection of a mode, it appears, to the person who may construct the car; or else knew of no mode, I should think.

" The wheels of his car are too close together to be drawn well by the body; their proximity would make the truck unsteady, if left free to run by the center pivot; and the mode of connecting the wheels, by springs, particlarly described by him, is not safe nor practical for railroad purposes. It would be impossible to keep such a construction of truck square on the curves, or the axis parallel, or prevent violent oscillation, and hence it would be too unsafe and unsteady. I would not assume the responsibility of putting it on the road, on any account, owing to the strong probability that loss of life would ensue from it, and it never has been either constructed for or in use on the Boston and Lowell Railroad, under either eight-wheeled passenger or freight car bodies.

" Placing the wheels in each truck closer together, or as close as possible, without the flanches touching, and placing the trucks further apart, is no invention; it is simply a change of proportion, and not a change or discovery of principle. The former makes the truck unsteady, and is no improvement; and the distance of the trucks apart, will follow or depend upon the increase or diminution of the length of the body, without effecting or altering the action or principle of the truck or car. As for example—diminishing the length of the body of the Tredgold car " A," would diminish the distance of the trucks apart; and putting a longer body A 2, on the same trucks, in the same essential manner, increases the distance between the trucks, but does not alter the principle or action of the car."

The next witness is Stephen W. Worden, defendants' No. 2, pages 39, 40—folios 165 to 168, where the practical results of different constructions are stated.

He says: "I have examined the specification of Ross Winans' patent, of Oct. 1st, 1834, and understand it. The model B, now before me, with a piece of paper on the truck, signed by Stephen Ustick, Wm. B. Aitkin, Oliver Byrne, George Beach and E. Martin, and on which I place my signature, is a true representation of the car described by the said specification of Ross Winans. Yesterday was the first time I ever saw or heard of this model.

"In the summer of 1838, I was superintendent of construction and repairs of cars for the Richmond, Fredericksburg and Potomac railroad. The shop was in Richmond, Va., and at that time a pair of trucks, constructed precisely similar to that model, and to the specification of Ross Winans' patent, were brought to me at the shop, and I had orders to put them under a car and try them.

"I understood that they were constructed in and brought from Baltimore. I put them under a *platform car*, what we called then a 'timber car,' and the car was then put in a train for trial, and on trial it proved a total failure. The trucks would not stay on the track, the car ran off, and the trucks were smashed to pieces, so much so that we considered them a total failure, and threw them away, as not practical and too dangerous to use. I afterwards put other trucks under the platform, the same as we had previously been using, that is, with a rigid frame like the Tredgold model, and the cars now in use. The Winans' truck, as described in his specification, is utterly worthless, and of no practical use. It is dangerous—too dangerous to use. I have no interest in this suit, either directly or indirectly."

There it seems that Winans' plan had been tested, and it was found that it would not stay on the track.

Then there is Henry Shultz, defendants' No. 4, pages 40, 43, 44—folios 154, 165, 166, 167, 168, 169.

He says, at folio 154: "I am a railroad car builder by occupation, and have been since the early part of the spring of the year 1832, at which time I commenced to work for the Baltimore and Ohio Railroad Company."

Then at folio 165, he continues: "In the year 1837 I engaged in car building with the Philadelphia, Wilmington and Baltimore Railroad Company, at Baltimore, and have had charge of the building and repairing here ever since. I have built nearly all of the eight-wheel passenger cars that the Philadelphia, Wilmington and Baltimore Railroad Company use this side of the Susquehanna river; and I understand the principles of construction and action of the eight-wheeled cars in general use.

"The eight-wheeled passenger cars in use have swinging bolsters and draw springs. The trucks are rigid, rectangular wheel frames, made as rigid and stiff as possible; and the cross bolsters and end pieces are secured by joint bolts. The distance of the body bolster from the end of the body brace, is about 7 feet; and the distance from the end of the box part about 4½ feet; and from 35 to 44 feet in length. The distance between the centres or bearing points of the wheels, in each truck on the rails, is made as far apart as we can get them in a truck from 7½ to 8 feet long; they are further apart than the wheels of the Columbus were; and to get them far apart is an advantage; it makes the car run easier and steadier than it would if they were closer together. The trucks have male and female transom plates, to connect them to the body and allow them to swivel, with a body bolt passing through them to hold them together. The male and female transom plates, for the trucks to swivel, are in universal use on the eight-wheel cars on railroads, and with swinging bolsters on the passenger cars; I have not seen any without them. I have examined and understand Ross

Winans' specification for an eight-wheel car, dated Oct. 1, 1834; at least, I understand what Ross Winans means to describe by it; but it will not answer for an eight-wheel car. A single pair of wheels, as he states, would turn right around into the middle of the track; and putting the wheels close together, as he describes, would tend to do the same thing. The springs that he describes for coupling the wheels, are good for nothing; they have been tried on the Baltimore and Ohio road, and condemned; they will not hold the wheels steady; and if one should break, it would destroy the whole truck. Coupling the truck with the body by a king bolt, like a common wagon, which he describes, is not the way in which eight-wheel cars are built. They have male and female transom plates; and the body bolt may be taken out, and yet the car will run. The body bolt is a security only in the eight-wheel cars; the swivelling is in the transom plates. There is no attempt in Mr. Winans' specification, to describe any mode of drawing or moving the car. The principles, or conditions and construction there set out in that specification, are unsafe, liable to accidents, and would not be suitable for railroad companies to use. I never saw his specification before, and have no interest whatever in the subject of this suit."

The next witness is Jacob Shryack, defendants' No. 3, pages 39, 40—folios 156, 157, 158, 159.

This witness testifies to the trial of the trucks, according to Winans' specification; and also in relation to the Washington cars. He is foreman of the building of passenger cars on the Baltimore and Ohio railroad.

At folio 156, he states: "On the 28th of October, 1834, I went into the employment of the Baltimore and Ohio railroad company, and have been in their employ ever since.

"I assisted in building the passenger cars of the Washington Branch road, that was opened in the latter part of the summer of 1835. We commenced to build the Washington cars in the early part of November, 1834, as soon as the benches and shop were prepared for the hands to work on them.

"They were built under the superintendence and direction of George Gillingham.

"There were twelve of the Washington cars built in 1835, at the Charles-street car shop. I think the trucks were built by Jacob Rupp. He was also building house cars at the time; that is, double pitch roofed freight cars; he built a great number of them. The Washington cars, built in 1834 and 1835, were not like the model B, now before me, and on the paper of which I have written my name. They had a wheel frame, made of timber and iron braces, that held the axles parallel to each other; and the action of the wheels was normal to the track. They were different altogether from the model B. The trucks on this model B are known as the spring trucks, and have been tried with strong springs, and a saddle of cast iron upon the top of each spring, with checks running along the sides of the spring to keep it from spreading; the bolster resting upon, and secured to the top of the saddle. The springs of this model have no saddle to keep them from spreading. These spring trucks, even when provided with the additional security of a saddle, have proved to be very poor. We can never keep them square, nor with the axles parallel. It is a dangerous thing, and is condemned as unfit for use on the Baltimore and Ohio road, where they have been tried. They were made under the administration of James Murray, as superintendent on the road. The model B is correct to the description in Winans' specification of October 1, 1834. A car like that is not practically useful, and not safe to run on a railroad."

This affidavit shows, that they were tried on the Baltimore and Ohio Railroad, with an improvement added to them, and yet they were good for nothing, and condemned.

This is the third witness on that point of construction.

I will now give your Honor the names of some additional witnesses, with references to their testimony, to save the time of the Court, and to relieve you from hearing their evidence read, as I have already collated most of the facts on this head of construction.

John B. Winslow, defendants' No. 3, pages 120, 121—fols. 479, 480, 481, 482, 483; page 122—fol. 488.

Asahel Durgan, defendants' No. 3, pages 123, 124—fols. 490, 491, 492, 493, 494, 495, 496.

John Crombie, defendants' No. 3, pages, 125, 126, 127, 128—folios 500, 501, 502, 503, 504, 508, 509, 510.

Henry W. Farley, defendants' No. 3, pages 128, 129, 130 —folios 512, 513, 514, 515, 516; page 131—folios 521, 522, 523.

Wm. P. Parrott, defendants' No. 3, pages 152, 153, 154— folios 608, 609, 610, 611, 612, 613, 614.

Septimus Norris, defendants' No. 4, pages 1, 2, 3—folios 1, 2, 3, 4, 5, 7, 8, 9, 10.

Mr. Norris is not only a builder of some of the best locomotives, but he is one of the best engineers in the United States.

M. W. Baldwin, defendants' No. 4, pages 3, 4, 5—folios 11, 12, 13, 14, 15, 16.

He also is one of the oldest builders of locomotives in the country, and one of the best engineers. His manufactory, as well as that of Mr. Norris, is in Philadelphia.

Richard French, defendants' No. 4, pages 5, 6, 7—folios 17, 18, 19, 20, 21, 22, 23, 25, 26.

There is but one more witness of the third class, who is a very intelligent man, whose testimony I will read on that point. He is Asahel Durgan, defendants' No. 3, pages 123, 124—folios 490, 491, 492, 493, 494, 495, 496, 497.

He says: "I am 42 years of age; reside at East Boston, and am a mechanic. I have been engaged on the Eastern Railroad for fourteen years as a master mechanic, and for the last four years have had charge of the shop for the construction and repairs of cars. I have often had occasion to observe the action of the eight-wheel cars in running on the road, to be able properly to repair the wear that takes place, and I understand the principle of construction and operation of the eight-wheel cars in use on railroads generally. In the construction and use of eight-wheel cars, it is all important that the trucks should be made and kept rigidly square, to hold the axles parallel and keep the wheels square on the track, and in order to secure these conditions, the wheel frames are braced as strongly as possible. On a road having sharp curves, the bearing points of the wheels on the rails in each truck should be secured a distance apart, the same as the guage or breadth of the track, and a few inches further apart than the guage of the track, if the curves are not very sharp. This relative distance causes the trucks to run steadily, and with the least friction of the flanches against the sides of the rails, both on the curves and the straight track, as it keeps the axles of the wheels very nearly parallel with the mean radial line of the curve between them, and the extent of the oscillation of the truck, when it passes over inequalities of level, is then the least, and the least injurious in its effects on the flanches, the wheel frame and the rails. The latitude between the flanches and the rails, necessarily allowed on the roads for the trucks to move forward easily, is about an inch. The resistance and friction that causes the trucks to turn or oscillate on the tracks, from inequalities of level, on the curves and straight track, at the points of rails, and from the different directions of the rails of the curves in their tendency to spring the trucks out of square, necessarily wears the flanches, and to a slight degree deranges the square of the truck, for I find the cars in this condition brought to the repair shop, to have the running gear repaired. In repairing cars brought to the shop, when I find the flanches much worn, as I do very often, I invariably place my squaring rule on the truck frame, to see if it is slightly strained out of square, and always find it to be so, and a cause of the running gear needing repair; when I always bring the truck back to its square and rebrace it. Preserving the rigid and square condition of the trucks, is necessary to the safety and good running of the car, and to prevent excessive wearing of the flanches. I have read and understand the specification of Ross Winans' patent, of October 1st, 1834, for an improvement in cars or carriages intended to run on railroads. The only truck that he describes, is made by connecting strong springs to the journal boxes of the wheel axles, and connecting the springs by a bolster, with centre pivots, connecting two of these trucks to the body of the car. The model B, signed by me, is such a car as I would make from the description of the patent. It is an accurate representation of the patented car. It would be impossible, in my opinion, to hold Ross Winans' truck square by the bolster and the springs mentioned and recommended by him, connecting the two axles of his trucks, and as the wheels are to be as stated by him, very close together, the consequent violent oscillation and friction produced, would twist the axles out of parallelism, and the wheels out of square, so much as to make it extremely unsafe—too unsafe for use. It would hop off the track, and if the spring broke, the side of the car would fall, and the destruction necessarily be considerable.

" The old plan of the rigid wheel frame, with the bearing points of the wheels on the rail, a distance apart equal to the guage of the track, described by Chapman in 1814, and Tredgold in 1825, organized, forming an eight-wheel car, the same in principle as those now generally used, is far safer and more pratical than Ross Winans' patented car."

The witnesses to whom I have referred, are not experts who make a business of testifying in courts. They are practical men and engineers, testifying in relation to matters which it is their business to understand, and hence there is no class of witnesses who are entitled to so much weight in the investigation of matters such as these. It has been necessary to multiply them in order to embrace all the points raised, and I believe that the principle, and the construction, and the purpose, has been covered by the testimony which I have adduced.

It is proved then, that to keep the wheels in each truck in line with the rails, as near as may be, keeps their axis near parallel with the radius of the curve, and the construction of a rigid rectangular frame, is necessary to effect this, in conjunction with the distance of the bearing points of the wheels apart on the rails, as near as may be, equal to the guage of the track.

This distance of the bearing points of the wheels, both before and behind the pivot, or centre bolt, gives a length of leverage, which safely controls the truck between the rails, and prevents it from turning the axis of the wheels too much across the radius of the curve of the road.

When the bearing points of the wheels on the rails are equally distant apart, or, as near so as may be, both across the track and in the length of the rails, the action of the forward motion of the wheels upon irregularities at the joints being incessant, tends continually to arrest the wheel and lurch the truck around; and this action is then, with such arrangement, counteracted and balanced by the re-action of the side of the rails against the flanges, without multiplying or increasing the force of the bearing of the flange against the side of the rail, above the measure of force of the action, causing the lurch of the truck.

Because, *as action and re-action are* EQUAL, as a law of force and the leverage across the tract is equal to the leverage along the rails, the mean line of force tends to intersect the king-bolt, and thus a BALANCE of force is exerted to keep the truck on the track, without increasing the force of the flanges against the sides of the rails, and hence, without increasing the tendency of the truck to catch and run over the rail.

All the best engineers therefore agree, that the true construction of the truck, is with a rigid wheel frame, holding the bearing points of the wheels equally distant and square on the track.

Suppose, for example, that a truck has a rigid unyielding frame, and the wheels with their bearing points equally distant on and across the rails, and a wheel strikes a joint of the rails, a stone, or any elevation on the track, with a measure of force equal to one thousand pounds. There is then a force of one thousand pounds tending to press the truck around on its centre off the the track, which brings the front flange in contact with the rail with a force of one thousand pounds.

The two balance each other, and the mean line of force is directed to the centre pin, which it intersects, and the truck is balanced without an increase in the force of the flange against the rail, beyond the force of the action which caused it. The danger of running off, increases with an increase of the force with which the flange is brought against the side of the rail, and the danger diminishes with a diminution of the force with which the flange is brought in contract with the rail.

The leverage about the king-bolt across the track, is *constant*. In a track of four feet eight inches and a half guage, and a lateral play of one and a half inch between the flanges and the rails, leaves four feet seven inches, and the half of this; two feet three and a half inches, is the constant leverage ACROSS the track around the king-bolt; it is the action of the wheels against the projections of the rails, that is exerted on this constant leverage.

Now, the leverage around the king bolt, on which the reaction operates when the wheels have struck a projection on the surface of the rail, is always equal to half the distance between the bearing points of the flanges against the side of the rail; the distance being taken in the direction of the rails, because the king bolt occupies a position half way between the bearing points. This may be a variable—not a constant—distance; because it may be diminished or increased by diminishing or increasing the distance between the centres, or bearing points of the wheels. If the wheels are brought closer together than a square, this leverage of the reaction is thereby diminished, and the flange bears with more force against the side of the rail.

Suppose that it is diminished from a square, say from 4 feet 7 inches, between the bearing points on each side, to 33 inches, with the thirty-inch wheels, used on the Baltimore and Ohio road—this is about Winans' distance—and the thousand pounds action is exerted to turn the truck. The flange will be brought against the side of the rail with an increased force, proportionate to a thousand pounds—as 4 feet 7 or 56 inches, are to 33 inches; which would be with a force of 1,696 pounds, tending to cause the wheel flange to catch, and run over the rail—as a measure of force, 696 pounds more than the measure of force which caused it.

But, by diminishing the distance between the bearing points on each side below the guage across the track, not only is the measure of force—of the lateral bearing of the flange against the rails, from action or obstructions on the surface—increased, but the obliquity of the truck in turning, until it has consumed or exhausted the lateral play between the flanges and the sides of the rails, is increased.

This increase of obliquity, causes the flange to bear more directly or obtusely against the rail, and, with its accompaniment, a diminished leverage before and behind the king bolt, greatly increases the danger of running over the rail, by thus bringing the bearing points, or wheels, as close together as possible on each side of the truck, and shows the fallacy of such a change.

The action which takes place on a CURVE, is still more decisive in showing the rigid frame and wheels, distant between the bearing points about the same as the guage of the track, to be the true construction and proper arrangement of the wheels in the trucks; and Winans' *not* to be the proper construction *nor* arrangement,

In running on a curve, the outer front wheel, owing to the tendency of the car or body to continue on a straight course, brings its flange violently in contact with the outer rail of the curve. If the truck-frame were of elastic material—such as a pair of long springs, constructed and applied as described by Winans, capable of accumulating a reactive force, to throw the wheel against the side of, and over, the rail, as the wheel pressed against it, and forced the spring on its elastic strain—this cumulative, reactive force of the spring would become exceedingly dangerous. In its struggles to force the wheel, which has its flange already bearing against the side of the rail, it would, on the first opportunity—for the level of the rail, or a break or joint in the rail, to allow the wheel to catch—spring it over the rail; and thus, as is technically called, make the wheel HOP over the rail—off of the track.

But the rigid frame has no such tendency. It possesses no elastic, or cumulative, reactive, transverse force; and hence, the wheel has nothing to tend to make it climb the rail, or cross over, other than the one primitive law of motion; that the body is tending to move in a straight line, and the frame is made so stiff and strong as to resist this force without twisting out of square.

The action produced by this law of motion, on the front outer wheel, causing it to bear, with the flanges, against the side of the rail, tends to retard the motion of that side of the truck, while the inner side of the truck, having no such restraint, and running on the shortest or inner rail of the curve, tends to crowd ahead.

Here, the great IMPORTANCE of having a frame to hold the wheels square, and prevent the front one from being pressed sideways, and the inner one ahead, and thus twisting the truck out of shape, is APPARENT.

This can *only* be prevented by having a rigid wheel frame, and *avoiding long springs*, or other elastic material, which will allow this twisting and distortion of the truck to take place.

Here, also, on a curve, is apparent the importance of keeping the bearing points of the wheels apart, to prevent the short swivelling, or obliquity, that would take place, like a single pair of wheels on a centre point, and cause too hard a bearing of the wheel against the outer rail, as well as throw the axis of the wheels across the radius of the curve, instead of keeping them parallel with it.

Practical men, and engineers of any experience and merit, all agree, that the true construction of the truck, and arrangement of the wheels in the truck of an eight-wheel car, is to have a rigid wheel frame, holding the wheels square on the track, and the bearing points of the wheels, as near as may be, a distance apart equal to the guage of the track; and, then, the application of the springs to the trucks must be by pedestals, to confine their elastic action, normal or perpendicular, to the track, and not such as with long springs, and not such as to allow them to accumulate reactive transverse force, or play, or HOP the wheels over the side of the rail.

A construction of truck, possessing or admitting of cumulative, reactive, or elastic transverse force, is worthless, and extremely dangerous, and unfit for use on a railroad.

The trucks of the eight-wheel cars are, and should be, constructed of wood, as it will not granulate or become oxydized, as is the case with iron, when caused to violently tremble.

Wood also possesses the most strength, as compared with its weight, as well as the greatest durability.

The wheel frames of the trucks are therefore constructed of wood.

The trucks of engines are constructed of wrought iron, because it occupies less space, where wood would interfere with the steam machinery, which often contracts the trucks too much. They are, however, steadied and directed by the axles and conical action of the driving wheels. The Norris's have lately re-arranged the steam gearing, and build their engine trucks square, as the proper construction.

The means by which the truck is to swivel or turn, is the next object of inquiry in the necessary construction, and this is effected in the eight-wheel cars by means of transom plates. Plain transom plates were used on the first eight-wheel car in use in this country, in 1829.

The Quincy car, together with side-bearings, and also in the wood and trussel cars, on the Baltimore and Ohio Railroad in 1830.

The transom plates now in use are male and female circular plates; when small in diameter, side bearings are required to steady the body of the car.

When the transom plates are of large diameter, from sixteen to twenty-four inches, side bearings are dispensed with.

These were first used on the model and the car "Victory," at Philadelphia, in 1834, 1835. To make the male and female transom plates move safe, a king bolt passes through the centre. With plain transom plates, like the Quincy car, a king-bolt is necessary, and is used to form the centre of the swiveling action of the truck.

The position of the trucks, in relation to the ends of the body, is another feature of the construction.

The eight-wheel car or carriage has swivelling running gear, and therefore it must be compared with cars or carriages of its kind; that is, with cars or carriages having swivelling running gear; and wagons, cars and carriages on the common road, turnpike road, and on railroads, with swivelling running gear, whether with two wheels or with four wheels, swivelling on a pivot, have always had the swivelling running gear placed under the ends of the body.

Also, it is a common elementary feature of mechanics, that every body of an eight-wheeled carriage or car is a trussel, which word in mechanics means a frame work resting with its ends on two separate points or objects; such as a trussel bridge, for example, which is a frame work resting with its ends on two abutments. And also common practice, as well as common sense, dictate that a carriage having swivelling running gear, whether of two or four wheels, or eight wheels, must rest with its END on the swivelling point. Common road wagons, carriages, railroad cars and other vehicles of common use, have been generally constructed with the END resting on the swivelling running gear.

Practice with the eight-wheel car on railroads, has established, that the truck must *not* be so near the end of the car, as swivelling running gear is commonly placed in wagons or carriages; that is, it must not be at or near the end, as stated by Winans.

In the eight-wheel car, the centre of the truck is placed about *seven feet from the end* of the bottom of the body framing, or platform which receives the concussion, when cars run together in trains, or in line on the road.

The truck, therefore, by practice, is established as most convenient under all circumstances, when distant from the ends of the bottom framing of the body, to the centre of the bolts or cross piece about seven feet; and this is the distance employed by the defendants, and in general use.

This distance protects the trucks from collision with each other, on the road and in trains; gives the body a fair balance of its weight over the trucks, consistent with its strength, to support the middle of the car, and divides the car in its support generally into about sixths, with a forty-two feet length of car.

One-sixth, that is seven feet, balanced on each side of each bolster, two-sixths or fourteen feet, to be sustained by the main strength of the body.

This is the general construction; some few cars, however, are sixty feet long, and some much shorter than forty-two feet.

It is established in this case, by the best of proof, and, in a measure, shown in this argument, that the construction or machinery necessary to embody the principle of an eight wheel car to use, consists of two four-wheel trucks, with stiff, rigid, rectangular wheel frames of wood, holding the four wheels on the track, with their bearing points on the rails, a distance apart, as near as may be, equal to the guage of the track. The springs, if applied between the wheels and frame, having pedestals or other fixture to confine their cumulative force, or elastic action, perpendicular or normal, to the surface of the rails, and these trucks supporting the body of the car, about seven feet from the ends of the bottom framing, connected by king-bolts, with transom plates of large size, or else of small size with side bearings.

The purpose effected by this construction is, to cause the car to run steadily and safely on the track.

The principle is, as before stated—the swivelling or turning MOTION of two four-wheel trucks, under one body, burthen or load; and such construction practically develops this principle, with the beneficial purpose, or result stated.

As to the mode of drawing the car, it may when run in trains, or when liable to come in collision with other cars, be drawn by a short perch from the truck; so short a perch, as to not project beyond the ends of the body, or by a chain from the front of the truck, or by a rod and chain, coupling from the centre pivot, or by a coupling from the middle of the end of the body. In all of which cases, when the cars run against each other, or back each other, they will strike and push by the body, and not by the truck or perch.

All of these are used at the present day.

When, however, the car is intended to run singly, the perch may project from the truck, beyond the end of the body, without serious objection, *such as in the Columbus.*

The majority of cars, however, both four-wheel and eight-wheel, are drawn by a coupling, from the middle of the end of the body; and this is the oldest method of drawing railroad cars, in trains, shown by the works or publications on railroads.

These different modes of drawing the car do not change its principle, though they may possess different degrees of merit, as modes of drawing.

Such is the principle, construction and purpose of the eight-wheel car, in general use, and constructed by the defendants.

The argument, also, in order to show most clearly what is the construction, to develop the principle, and effect the purpose, has shown, we submit, that connecting the axles by long springs, the wheels closely, and at or near the ends, as described by Winans, is *not* the construction to develop the principle; it is one to be avoided by the car builder and railroad manager.

The next subject that I will examine is involved in the inquiry as to

What is the origin and history of this eight-wheel car? and

What improvements, for convenience and comfort, have lately been made upon it?

I will at this time mention the improvements made upon it.

They are, Davenport and Bridges' patent swinging or pendulum bolster, in general use; and Condue Gatch's changeable backs, with swivel joints, for the seats inside;

a new draw-spring coupling, for which a patent has been obtained; Stevens' improved equalizing brakes; and Kite's patent safety beams to the trucks.

These are all the essential improvements to be found in them, as differing from their original construction prior to 1831.

I will now proceed to examine the origin of the car.

We first find the principle of the car, and also the necessary construction or mechanism to develop the principle, in the Chapman specification, in vol. 24th of the "Repertory of Arts," printed and published in 1814, in London, a period of twenty years before the date of the plaintiff's patent, on page 130–139, with a general reference to the specification.

The specification is divided into different parts, with different figures relating to the different parts.

The portions of the specifications we refer to are in these words:

"We also, as the carriage containing the motive power will, thus loaded, be too heavy in various cases for the strength of the existing iron or wooden rails, if resting on four wheels only, so arrange it for such ways (or other confined ways, where the ledges either of the ways or of the wheels regulate the direction of the carriage) that it may rest equably, and move freely round curves or angles, either on six or eight wheels, so as to reduce its pressure on each in the inverse proportion of its number of wheels. Having thus described the outlines of the separate leading parts of our invention, we shall proceed to the means of carrying them into effect."

"Fig. viii. shows a carriage of six wheels for the engine, which may rest equably, or nearly so, on each of its wheels, and move freely round the curves or past the angles of a railway. 1, 1, the fore pair of wheels, are, as usual on railways, fixed to the body of the carriage; 2, 2, and 3, 3, the other two pair, are fixed (on axles parallel to each other) to a separate frame, over which the body of the carriage should be so poised as that two-thirds of its weight should lie over the central point of the four wheels, where the pivot 4 is placed, and the remaining third over the axis 1, 1. The two-thirds weight of the carriage should rest on conical wheels or rollers, bearing upon the curved plates *c, c*, so as to admit the ledges of the wheels, or those of the way, to guide them on its curves or past its angles, by forcing the transom or frame to turn on the pivot, and thus arrange the wheels to the course of the way, similarly to the carriage of a coal wagon. And if the weight of the locomotive engine should require eight wheels, it is only requisite to substitute, in place of the axis 1, 1, a transom, such as described (laying the weight equably upon both), and then similarly to two coal wagons attached together, the whole four pair of wheels will arrange themselves to the curves of the railway."

This Chapman specification describes both a six-wheel and an eight-wheel carriage. It distinctly, also, speaks of ledges or flanges on the wheels. The six-wheel carriage has a truck frame, with four wheels, and a single pair of wheels on a fixed, revolving axle. Plaintiff's attorney stated that the wheels revolved on the axles. The specification contradicts this; for it expressly states that the wheels are "*fixed* on axles parallel to each other;" therefore, the axles must revolve with the wheels.

This construction of running gear, in connection with the body, is now in use on steam carriages, and on carriages containing wood, water, and such compact and heavy articles, requiring but little comparative room.

The eight-wheel carriage is now in general use for carrying bulky articles and passengers, and is the freight and passenger car of the present day.

Few of the carriages on railroads contain the motive power as well as passengers or freight, as was the case with steam coaches and carriages in Europe at an early day, and as mentioned by Chapman.

The model *K*, in this case, is proved by the most competent witnesses on the subject in the country, to be a true representation of the Chapman carriages—both the six and eight-wheeled.

" Any motive power," such as Chapman mentions, whether horse power, man power, or steam power, may be put in it; or passengers or freight may be put in it; and it may be drawn by any of the modern locomotives, or horses, at the highest speed.

The change in the speed, or efficiency of the means of drawing the car, does not change the arrangement nor principle in the running gear, and the connection of the same with the body. We deny that the arrangement or principle reposes in, or is governed by, the speed.

Chapman speaks of its being drawn by a chain, which is connected to the body of the car, and derives motion from a barrel or windlass ahead, on the track.

The modern locomotive is an improved machine for drawing this car; and the entire capacity, or space, in the body of the carriage, is now appropriated to the stowage of goods or the reception of seats for passengers.

The steam passenger coach, &c., is now again being tried in New-England, with the engine, and room for passengers in the same vehicle.

The essential principles by which the carriage, on its eight wheels, conforms to the curves, and pursues a steady and safe course on the road, remains unchanged.

This Chapman carriage, on the eight wheels, has several purposes and objects to attain.

It has eight wheels, to distribute the great weight on more than four or six bearing points, and properly on the road, essentially as now practiced in the eight-wheel car. But it is observable, from the specification and the drawing, that those eight wheels are not put in a single rigid frame; because other objects are to be attained, besides the distribution of the weight among the numerous wheels, and properly on the rail.

It has eight wheels to distribute the weight; but those eight wheels are divided into two sets of four wheels each, independent of each other, with wheel-frames, king-bolts, transom-centres, and side-bearings,—for what?

To distribute the weight? No, sir.

These two trucks would not be necessary to distribute the weight among numerous wheels, and properly on the rails.

A single stiff frame, with eight-wheels, would do that, though not *equally* among the wheels.

They are put in two trucks of four-wheels each, with king-bolts, transoms, and side-bearings, that the two trucks shall have a swivelling motion, under the one body of the carriage, *to turn curves and run safely and steadily on the road,* and so the printed publication states.

Here, then, this Chapman carriage embodies the whole principle and purpose of the eight-wheel car, as made by the defendants, and in general use.

The next point in this car is, as to the construction of the trucks, and their arrangement with the body of the carriage.

The truck, or bearing carriage, is described in the Chapman specification and drawing, and fully proved to be composed of a rigid, rectangular wheel frame, consisting of two side, and three cross-pieces.

The middle cross-piece carries the transom, centre-plate, and king-bolt, and side-bearings, and sustains the weight of the body carried.

The body has anti-friction, side bearing rollers, to allow the trucks to swivel easily.

The axles of the wheels are described as parallel; and the construction of the truck frame, by its rigidity, necessarily holds the axis parallel.

The object of the transom or truck, is expressed as *to turn curves.*

The drawing of the truck or transom, given in the specification, measures from the centre of one wheel to the centre of the other, on each side, *the same in distance, as the guage of the track shown in the plan.*

The centre or king-bolt, is *equi-distant* from the bearing points of the wheels.

The *wheels*, therefore, are held by the truck frame, *square on the track*,

The two trucks or transoms, for an eight-wheel car, are to sustain the *one carriage body*. They are described and shown as *swivelling* running gear. All swivelling running gear, in the common road wagons, carriages, and all sorts of vehicles, has been, from time immemorial, placed under the ends of the body.

The repetition of this fact was not necessary for Chapman to make.

All mechanics knew where swivelling running gear was placed; and so these four-wheel transoms would, of course, be placed where swivelling running gear had *always been placed*, i. e., *under the ends of the body.*

In fact, a man ought to be considered an idiot, or excessively ignorant, who would suggest to place it anywhere else than under the ends of the body; for he is not directed by Chapman to make any change, in this respect, from other swivelling vehicles, whether used on common, turnpike, or rail roads, and could give no sensible reason why a change should be made, or why the swivelling gear should be placed elsewhere than under the ends of the body.

To support the body equably, also, as Chapman describes, requires them to be placed under the ends of the body.

All the essential differences between the eight-wheel car and other vehicles, whether of four or six wheels, are described in Chapman's specification; and the principle and precise essential construction and mode of connecting and arranging the wheels, as used in the eight-wheel cars generally, and constructed by these defendants, are there described and fully explained.

As to the purpose or effect, we have shown that the principle of the Chapman carriage is the same, and that the essential construction to develop the principle to use, is the same as now used. The purpose is expressed to turn curves and run on a railroad; and as to the effect, it is an immutable principle in physics, displaying the stability of creation and the character of physical laws, that "*like causes produce* like effects;" while, to conclude this principle of physics, I will remark that it is equally true, as a physical evidence of the omnipotence and omniscience of the Creator, that the same effects may be produced by different *physical laws* and *properties of matter*, or, legally expressed, by *substantially* different causes. If the *knowledge* to create physical laws to produce a certain effect were *limited* with the Creator, there would be *no omniscience*. If the *power* to create those physical laws were *limited*, then there would be *no omnipotence*. Hence the truth in physics, as I have stated, "that the same effects may be produced by different principles or physical laws," not only exists, but is a striking evidence of omniscience and omnipotence, supporting our belief in this truth of law and fact.

We have shown the principle, and the construction or mechanism to embody the principle to use, to be the same as now in general use, and built by these defendants; it therefore follows, of course, by the laws impressed on matter, that the car must run, as near as may be, with the same safety and steadiness on the road.

Additional wheels have been made for the model *K* of the Chapman carriage, to be placed on the same axles, instead of the smaller wheels, and practically show that, while such change of the size of wheel brings the flanges nearer together, the essential parts of the trucks, and the bearing points on the rails, remain the same distance apart. The large wheels represent about thirty-six inches in diameter; the small ones, about twenty inches diameter. The large size, or from twenty-eight to thirty-six inches, are now more generally in use on eight-wheel cars, though they still vary on cars from twenty to thirty-six inches. This change in the size of the wheels, does not change the essential construction of the car, although it does bring the nearest faces of the wheels closer to each other, and of course the remote faces as much further from each other in the same truck.

The bearing points and axis remain the same distance apart.

It is through the bearing points of the wheels, on the rails, that the controlling action and reaction, which directs and keeps the car on the track, takes place. The wheels there, at the bearing points, act on the rails, and the rails react through these bearing points on the wheels; therefore, the distance between the bearing points of the wheels, in each truck, is the essential distance in the development of the principle of the car.

I will refer to some witnesses upon these points. These witnesses were not in the Troy and Schenectady case; and they are not men who, for a living, make a habit of appearing as experts. They are practical men, ornaments to the community, who must possess the best knowledge on the subject, and such as are most to be relied upon by the court, in cases where a principle of construction and mechanical action is the subject of investigation.

Wm. B. Aitken, defendants' No. 3, pages 131, 132—folios 525, 526, 527.

He says: "I am the same who, on the 19th day of July, 1852, made an affidavit in this case; that he is a pattern and model maker by trade and occupation, and understands the descriptions and drawings of machinery generally, as it is his business to work from them, and also, in many instances, to make drawings of new machinery, from which to make models and patterns. That on or about the middle of last week, Wm. W. Hubbell, Esq., handed to me the 24th volume, 2d series, of the Repertory of Arts, and requested me to make a correct model of the engine or carriage described in the specification and drawing of W. & E. W. Chapman, published therein in February, 1814, to show its principle for turning curves and running on a railroad; and also to make an entire set of wheels for it, of larger diameter than those shown in the drawings, fig. viii. I have made the said model exclusively from the said specification and drawing of Chapman, as contained in said book, to turn curves and the angles of a railway, and the same is marked K, and I have signed my name on a piece of paper attached to one of the trucks.

The proportions and construction of the model are correct to the description in the book. The extra large wheels are marked L."

Here is a practical man who makes the car itself, in its essential principle and construction, as now used from the specification in the book, by his own comprehension and understanding alone.

Another witness is Wm. J. McAlpine, defendants' No. 3, pages 93, 94—folios 372, 373, 374, 375.

He says: "Both the six and the eight-wheeled car or carriage, in all its most essential principles and details of construction and operation, to conform to curves and other inequalities of a railroad, and divide the pressure equally among the wheels, are handed down to us from England, in the printed publications, Repertory of Arts, Wood's Treatise, and Tredgold's Treatise. In the 24th vol. of the Repertory of Arts, 2d series, the Chapman specification, printed and published in Feb., 1814, in London, describes a six-wheeled carriage having a regular four-wheel frame, with centre-pivot and side-bearings under one end, and a pair of wheels under the other end of the carriage body. It is ascertained by measurement of the plan of this four-wheel truck, or wheel-frame of four wheels, in fig. viii, plate v, that the distance of the bearing points of its wheels, apart, is about equal to the breadth of the track; and the specification describes that the carriage body may be supported by two of these wheel frames, both alike, laying the weight equally upon both, and they turn by means of their centre pivots, or axis, to suit the curves and other inequalities of the roads. In that description and drawing of Chapman, we have the essential principle of construction and operation, both in detail and organization, of the eight-wheeled car now used. There is the rigid rectangular wheel-frame of four wheels, with the cross-piece, centre-pivot and side-bearings; and the bearing points of the wheels, whether larger or smaller wheels, may be used, given a distance apart, about the same as the breadth of the track, and the carriage body sustained by the two bearing carriages, laying the weight equally, being the same essential principles of construction as used at present in the eight-wheeled car.

George W. Smith, defendants' No. 3, pages 75, 76—folios 297, 298, 299, 300, 301. This gentleman is one of the founders of the Franklin Institute, of the State of Pennsylvania, for the promotion of mechanic arts, founded in 1824, &c., &c.

At folio 297, he says: "Also, the principle is sufficiently set forth to enable an ordinary mechanic to construct the same, in the specification of the Chapman patent of 1812, and printed and published in 1814 in the 24th volume, second series, of the Repertory of Arts, pages 130 and 139, and the drawing plate v.

"The necessity of keeping the axles parallel, and use of a rigid rectangular wheel frame, that will hold the wheels square, as at present used in the eight-wheeled cars, are there clearly set forth and shown. A ground plan of the wheel-frame of one truck is clearly shown in the engraving, fig. viii.; it has the two side-pieces and the three cross-pieces, one a middle-piece with the pivot-centre; being the same comprised by the Tredgold car, and in the eight-wheeled cars now in use on railroads. The distance between the axles of Chapman's wheel-frame measures about the same as the breadth of the track. The wheel appears small in diameter, which gives, apparently, more distance between them than if they were larger in diameter. The distance between the bearing points of these wheels on the rails, or the centres of their axles, which is the essential distance, being about the same as the breadth of the track, makes this wheel-frame, in fig. viii. of Chapman's, in all its essential features, the same as the Tredgold wheel-frames and those now in use on eight-wheeled cars."

Where it is desired, therefore, to have a more capacious body, the only change to be made is to lengthen the body, allowing the trucks to remain, as before, near the ends.

The means of connection with the body, and the distribution of the weight, remain the same; and the length of the body, whether shorter or longer, does not change the principle, nor the essential operation nor effect.

The next witness is Vincent Blackburn, defendants' No. 3, page 97—folios 385, 386, 387, 388.

He says: " In the construction of the wheel-frames or trucks, as they are commonly called in the United States, used on the eight-wheeled cars generally, the axis of the wheels are held parallel to each other by rigid rectangular wheel-frames, having centre-pivots, the same in principle as the wheel-frame or truck shown and described in fig. viii., plate v., in Chapman's specification, printed and published in the Repertory of Arts, volume 24th, 2d series, 1814, in which also the bearing points of the wheels are a distance apart, about the same as the breadth of the track ; and two of these trucks are described for an eight-wheel carriage, and one with a pair of wheels for a six-wheel carriage. Both of these plans are used on the Utica and Schenectady railroad. Some of the carriages, carrying the wood and water, commonly called the tender, have the front pair of wheels attached to the body of the carriage; and the four-wheel truck or wheel-frame sustains the back end of the body with side bearings and a pivot, the same as in Chapman's specification. Also, some of the tenders have two of these trucks or bearing carriages; and the freight and passenger cars have the two four wheel trucks, sustaining the body of the carriage, and constructed, in their essential details and principles, with the bearing points of the wheels about the same distance apart, as described and shown in the drawing and description in Chapman's specification, in the Repertory of Arts. He there describes and shows both the six and the eight-wheeled carriages, in their most essential principles, with rectangular wheel-frames, holdidg the axles parallel, and the bearing points of the wheels in the truck a distance apart about equal to the breadth of the track, with the usual cross-piece and centre-pivot to turn curves, the same as used on the Utica and Schenectady railroad ; and as to the eight-wheeled carriages, the same as generally in use on railroads."

The next is Charles B. Stuart, defendants' No. 3, page 91, folios 362, 363, 364.

He says: "I have examined the specification and drawing of a patent granted to W. Chapman, civil engineer, and E. Walton Chapman, of England, dated December 30th, 1812, as printed and published in the Repertory of Arts, Manufactures and Agriculture, vol. 24th, 2d series, Feb., 1814, published by J. Wyatt, London, England, and find there described a truck of four wheels, with rigid frame, and the bearing points of the wheels a proper distance apart on the rails, about equal to the breadth of the track, to be placed under the ends of railway carriages, one of them with a single pair of wheels, for a six-wheel carriage, similar to the six-wheel American locomotives, and two of these trucks for an eight-wheeled carriage, if desired, similar to the eight-wheeled carriages now used, for the purpose of allowing the carriage to move freely round curves, or pass the angles of a railway, allowing the frames to turn on a pivot, and distribute the weight eqally on the track, as is common in the cars now in use in the United States generally. An ordinary mechanic would be able to make the eight-wheeled car from that description. The details of construction and arrangement there shown are the true and practical principles of the eight-wheeled car. The description is much more practical and better in principle than the Winans' specification."

The next is George S. Griggs, defendants' No. 3, pages 105, 106—folios 419, 420, 421, 422, 423.

He says, at folio 419: "The whole of the essential construction and organization of the eight-wheeled car, as now used by the railroads generally, is contained in the specification and drawing of the six and eight-wheeled carriages of William and Edward Chapman, printed and published in the 24th vol., 2d series, of the Repertory of Arts, in the year 1814, in London. The drawing, fig. viii. of the specification, shows the transom or wheel-frame with its four wheels, having the centres of the wheels or their bearing points on the rails a distance apart equal to the breadth of the track; and the specification distinctly states that two of these, both alike, should be used for an eight-wheeled carriage, and one of them with a pair of wheels for a six-wheeled carriage. They are used both ways on the railroads in this country, and the model K, on the paper of which I have signed my name, is a correct representation of the carriage described in Chapman's specification in the Repertory of Arts. This model exemplifies both the six and the eight-wheeled carriage described by Chapman and now in use. The six-wheeled plan is used chiefly to carry wood and water for the engine, and the eight-wheeled plan is used for passenger cars and freight cars. All the essential principles of the distance of the bearing points of the wheels apart, and of constructing the rigid rectangular wheel-frame, to hold the wheels square and the axles parallel with the centre pivot, and the organizing of two of these with the same body to turn the curves and pursue a smooth, even, and safe course, the same as used on railroads, are described and shown in Chapman's specification, and in the model K which corresponds with it; and putting the larger wheels L, in place of the small wheels K, on the same axles, would bring the faces of the wheels closer together, but would not alter the principles; the essential proportion and construction would remain the same, whether horses or men are put in it to propel themselves, as stated by Chapman, or are in it, as freight or passengers; being drawn by a steam engine attached to it, makes no difference in the principle of the car; it still has always the eight wheels to distribute the pressure in two separate trucks to freely turn the curves; and these trucks are properly constructed and proportioned, and supporting one body, and containing all the essential construction and organization of the eight-wheeled cars now in use."

The next is Walter McQueen, defendants' No. 3, pages 100, 101—folios 398, 399, 400, 401.

After stating the principle, he says: "This is the essential arrangement and principle of the eight-wheeled car. The distance of the trucks apart depends upon the length of the car, and will vary with a variation of the length of the body, whether shorter or longer, without changing the essential principle. This essential principle of the eight-wheeled carriage, whether it has horses or men, or steam, in it to move or draw it, together with freight or passengers, or is used for freight or passengers alone, was given and is to be found described and shown in Chapman's specification, contained in the 24th volume, 2d series, of the Repertory of Arts, printed and published in London, in 1814. That specification and drawing, on pages 130 and 139, and fig. viii, plate v, distinctly describes and shows both the six-wheel and eight-wheeled carriage, in all the essential detail and organization or combination, as at the present time running on the Utica and Schenectady railroad, and also on the Albany and Schenectady railroad, to turn the curves, divide the pressure on the wheels, and run safely, smoothly and in a direct course on the track. I constructed some of the six-wheeled carriages for the company to carry wood and water; and the eight-wheeled carriages are the same as constructed by Eaton, Gilbert & Co., of Troy, and such as are used on railroads generally to convey freight of all kinds and passengers. The truck, or four-wheel frame, in fig. viii, and as described, has the rigid rectangular wheel-frame, holding the axis of the wheels parallel, and the wheels square on the track, and the distance between the bearing points of the wheels is about, or exactly the same, as the breadth of the track; has also the centre-pivot and side-bearings, and two of them, both alike, and in the same way support the carriage body under which they turn on the pivots to the curves and other inequalities, the same in arrangement and principle as the eight-wheeled carriages now used on railways."

The next witness is James H. Anderson, defendants' No. 3, pages 118, 119—folios 471, 472, 473.

He says: " The essential principle of construction and action of the eight-wheeled cars in general use, and manner of arranging and connecting the wheels with each other, and the car body as practiced commonly, is fully described in Chapman's specification, printed and published in the 24th volume of the Repertory of Arts, second series, published in London in 1814; and the six-wheeled carriage there described, and also the eight-wheeled carriage are now in use in this country.

"The model K, signed by me, represents them both being changeable. It is such a model as I should construct to represent the description exactly. I could build the eight-wheeled cars now in general use in all the essential construction, arrangement and organization, as I have described, from that description and drawing in the Repertory of Arts."

I will not read any further testimony relating to the Chapman car, as it will be, generally, a repetition in substance of that which I have already read; but I will refer to the remaining witnesses, giving their names, and the place where their evidence is to be found.

Asahel Durgan, defendants' No. 3, pages 124, 125—folios 496, 497, 498.

John B. Winslow, defendants' No. 3, page 121—folios 484, 485, 486, 487.

Henry W. Farley, defendants' No. 3, page 130—folios 517, 518.

John Crombie, defendants' No. 3, pages 126, 127—folios 504, 505, 506, 507, 508.

Wm. P. Parrott, defendants' No. 3, page 154—folios 614, 615, 616, 617.

Mr. Parrott is one of the most eminent engineers. The other gentlemen mentioned are engineers and practical men.

Wm. C. Young was educated at West Point, and is one of the best engineers in the United States; his testimony is in defendants' No. 3, page 151—folios 602, 603, 604.

Godfrey B. King, defendants' No. 4, pages 79, 80—folios 300, 301, 302.

Mr. King is a practical man, and one of the best machinists in the country; he has been engaged as master machinist for 17 years, on the Boston and Lowell railroad.

[Adjourned; and on the following day Mr. HUBBELL continued his argument.]

The next point to which I will advert, is — the manner in which the Chapman car was presented in the Troy case, with a view to show that the issue there raised on it was a different issue from that now raised.

On this point, I refer to the affidavit of Isaac Adams, pages 77, 78, 79.

The question in the Troy case was — whether the Chapman car did, or did not, embody the doctrines and mechanical structure of Winans' specification.

In this case, we say that the Chapman car embodies the principle of the eight-wheel car, and the essential structure to develop the principle, the same as that contained in the cars made by the defendants, to wit : to keep the axles parallel with the bearing points of the wheels in each truck — a distance apart, as near as may be, equal to the guage of the track, and held thus square, and in this position, by a rigid rectangular wheel-frame; two of which support the body, with an equal amount of weight on each, and have the swivelling motion under it, *to conform to the curvatures of the road;* while the increased number of bearing points distributes the weight over the road.

And further, we say, that Winans does not profess to have discovered the principle, in its true sense, which I have stated heretofore.

But he sets out the doctrine that a proximity of the wheels, as close as possible, that the axles may more nearly coincide with the same radius, like a single wheel, while the wheels thus brought together in each truck still remain in the two trucks remotely apart; and employing big springs and wagon bolsters, or analogous devices, to thus couple the axles and wheels as close as possible, without touching, and at the ends of the body, is his theory and mechanism, and that it tends to concentrate the weight.

We say that this is equally different from both the defendants' and Chapmans'.

And if his patent be construed so broadly as to cover the defendants', it necessarily must cover Chapmans', and is too broad, and void.

But if it be construed directly as to his theory of coincidence with the radius, by proximity as near as possible, and the mechanism is by spring connections, with common wagon bolsters, which he describes, then it is equally different from both the defendants' and Chapmans', and the defendants do not infringe.

To prove that this Chapman car was presented in the Troy case solely for the purpose of destroying the plaintiff's patent, I will refer to the charge of Judge Conkling, to the testimony of Edward S. Renwick, pages 80, 81, and to Harvey Waters, page 82, in defendants' copy of Troy case.

These were the defendants' witnesses.

Not one of them was a practical railroad engineer, and hence they were not the proper kind of witnesses. They were not such witnesses as we have now presented to the court.

Although they are very respectable gentlemen, yet they are not practical men, acquainted with the true principle and construction of eight-wheel cars.

The plaintiff's witnesses are Mr. Hibbard, pages 85, 86, 87, 88; and Walter R. Johnson, pages 91, 92, 93.

The court will understand that the point is : that in the Troy case, the Chapman car was put solely upon the ground that it was the same thing as Winans describes ; and we say that such was not the issue that should have been taken on it, for we put it on the ground that it is the same as the defendants', and that they are both equally different from Winans'.

Hence, if Winans' patent includes the one, it will include the other ; and if it includes Chapmans', his patent would be void ; but if it do not, then the defendants are not infringers. We do not understand, however, that Mr. Winans' specification, in the view that we have taken of it, and as expressed by the specification itself, includes such cars as those constructed by the defendants, nor such as Chapmans'.

I will read from Judge Conkling's charge to show the manner in which the Chapman car was presented in the Troy case. Page 131 of defendants' copy, and page 9 of the charge, in the Appendix to plaintiff's evidence.

The Judge says: " On the part of the defendants, however, it is insisted that carriages substantially like that in question were previously described in certain public works; and to establish this, they produce two English books, Wood on Railroads, published in 1825, and a volume of the Repertory of Arts, published in 1814. They insist that each of these contains a description and drawing of what is substantially like the railroad car which the plaintiff has patented. I shall not enter into a detailed explanation of the drawings and descriptions contained in these books, nor recapitulate the views and arguments concerning them, which have been so elaborately presented to your consideration by the counsel on the one side and the other, but shall content myself with calling your attention, presently, to some principles of law applicable to the subject."

We there see that it was insisted by the defendants that those cars were substantially like the railroad car which the plaintiff had patented, with a view to destroy the patent by showing want of novelty. These words of the Judge show that fact, viz: "They insist that each of these contains a description and drawing of what is substantially like the railroad car which the plaintiff has patented."

That is not the same case as we now present to this court. In that case, it was a different question entirely, and would depend upon the construction put by the court on Winans' patent. We think, however, that it will not bear such a construction as to include either the defendants' car, in this case, or the Chapman car. We contend that it is for a collective claim; i. e., for *the manner of connecting and arranging the wheels as described.*

The "arrangement" refers to the distance of the axles, which are to be coincident to the radius, as near as may be, in each truck, and the trucks as remote as possible; and the "connection" is by big springs, bolted to the boxes of the axles, and wagon bolsters across to the middles of the springs.

In this relation let me remark that a patent cannot be sustained for a change in distance between the parts, or the space between two objects in existing machinery for a similar purpose; nor for changes, whether chemical or mechanical, effected by *lapse of time*, in conjunction with the use of known machines, employed for an analogous purpose; nor for changes that take place by *natural* operations, or effects of existing machinery; nor for an idea; nor for a principle in the abstract.

And the Winans' patent, therefore, must be for the big springs bolted to the wheel boxes, with wagon bolsters between their middles, connected by king-bolts to a connection or body between them, or some analogous or elastic fixture, secured in like manner to obtain this proximity of the axles, or near coincidence to a radius, like the axle of a single pair of wheels; subject, however, to the question whether it is so pernicious as to be invalid.

We have shown that a car, full of passengers, with a single spring, moving at the rate of twenty miles an hour between them and eternity, is a position which the boldest man may well shrink from with fear, and is too dangerous to be of public use.

We say, that if the court construes Winans' patent to be for the manner of connecting the wheels with each other by means of the big spring, or an analogous device, as close as possible without the flanges touching—to embody the idea expressed in the specification of having the axles as near coincident with the radius, or like a single pair of wheels, and thus connected with each other, also with the body by the common wagon bolster and king-bolt, in the collective sense—then it is equally unlike Chapmans' and the defendants', and there is no infringement.

In order to show that such is his theory and mechanism, I will collate some parts of his specification.

First, I will examine those parts which set out his theory.

On page 111 of Appendix *A*, in plaintiff's proofs, folio 4, Ross Winans himself says: " From this consideration, when taken alone, it would appear to *be best to place the axles as near to each other as possible; thus causing them to approach more nearly to the direction of the* RADII *of the curves,* and the planes of the wheels to conform to the line of the rails."

I will now turn to page 113, folio 10, where he continues: " The two wheels on either side of the carriages are to be placed *very near to each other;* the *spaces between their flanges need be no greater than is necessary to prevent their contact with each other."*

He then goes on to the construction, at folio 14, page 114, and, after making a suggestion, at the commencement of the paragraph, as to construction, the theory begins with the word "provided," and goes on : " Provided that the fore and hind wheels of each of them be placed *very near together, because the closeness of the fore and hind wheels of each bearing carriage, taken in connection with the use of two bearing carriages coupled as remotely from each other as can conveniently be done* for the support of one body, with a *view to the objects* and *on the principles herein set forth,* is considered by me as a *most important* feature of my invention."

He says in his suggestion, as to construction : " Although I prefer the use of a single spring to a pair of wheels as above described, instead of the ordinary spring to each wheel, and consider it more simple, cheap and convenient than any other arrangement; the end which I have in view may nevertheless be obtained by constructing the bearing carriages in any of the modes usually practiced."

He then proceeds with his theory, as previously stated, with the proviso as to *proximity,* which he says " *is a most important feature of my invention"*

I next refer to folio 116, at the top of page 115, where he adds: " *The two wheels on either side of one of the bearing carriages may, from their proximity, be considered as acting like a single wheel."*

Now, to get at the sense direct, take the words that I have just quoted in connection with those to which I referred in the commencement of his theory, viz : " From this consideration, when taken alone, it would appear to be best to place the axles as near to each other as possible, thus causing them to approach more nearly to the direction of the radii of the curves, and the planes of the wheels to conform to the line of the rails," and you will discover that Winans' theory was *coincidence of the axles in each truck with the same radial line, as near as may be, like the axles of a single pair of wheels, and these trucks connected to the ends of the body.* That is the first branch to be considered and determined in an analysis of his specification.

Now, as to the mechanical *construction.* It will be perceived that Winans does *not profess to have discovered and applied* the " principle of the swivelling motion of two four-wheel trucks under one body, burthen or load."

His theory in each bearing carriage is coincidence of the axles to the radius, as near as possible, to act like a single pair of wheels, in connection with the two carriages coupled at or near the ends, a convenient distance apart. He does not profess to say that the two carriages have not been coupled near the ends as conveniently as may be, with the axles further apart, than with a design to have them coincident, as near as possible like a single wheel, and swivelling under the body in eight-wheel cars; his specification implies that they have been so constructed and arranged, and so operated.

His claim is collective. We have his theory or doctrine as to the arrangement, as to the manner of connecting and *arranging* the eight wheels, as described.

We have the arrangement already collated.

Now, what is the manner of connection, or the machinery described ?

At folio 9, page 113, of Appendix *A*, in plaintiff's proofs, he says: " For this purpose, I construct two bearing carriages, each with four wheels, which are to sustain the body of the passenger or other car, by placing one of them at or near each end of it, in a way to be presently described."

Then, in folio 10, page 113, he adds: " These wheels I connect together by means of a very strong spring, say double the usual strength employed for ordinary cars; the ends of which springs are bolted, or otherwise secured to the upper sides of the boxes, which rest on the journals of the axles; the longer leaves of the springs being placed downwards, and surmounted by the shorter leaves.

"Having thus connected two pairs of wheels together, I unite them into a four-wheel bearing carriage by means of their axles, and a bolster, of the proper length, extending across, between two pairs of wheels, from the centre of one spring to that of the other, and securely fastened to the tops of them. This bolster must be of sufficient strength to bear a load upon its centre of four or five tons. Upon this first bolster I place another of equal strength, and connect the two together by a centre pin, or bolt, passing down through them, and thus allowing them to swivel or turn upon each other, in the manner of the front bolster of a common road wagon.

"I prefer making these bolsters of wrought or cast iron. Wood, however, may be used.

"I prepare each of the bearing carriages in precisely the same way.

"The body of the passenger, or other car, I make of double the ordinary length of those which run on four wheels, and capable of carrying double their load.

"This body I place so as to rest its whole weight upon the two upper bolsters of the two before mentioned bearing carriages, or running gear."

Then follows some variable modes of suspending the body; after which, on page 114, folio 13, he continues:

"When the bolsters of the bearing carriages are placed under the extreme ends of the body, the relief from shocks and concussions, and from lateral vibrations, is greater than it is when the bolsters are placed between the middle and the ends of the body; and this relief is not materially varied by increasing or diminishing the length of the body, while the extreme ends of it continue to rest on the bolsters of the bearing cars, the load being supposed to be equally distributed over the entire length of the body."

92   ARGUMENT OF

Immediately following, is the suggestive passage which I have already read, but which I will repeat in this connection.

He says: "Although I prefer the use of a single spring to a pair of wheels, above described, instead of the ordinary spring to each wheel, and consider it as more simple, cheap and convenient than any other arrangement; the end which I have in view may, nevertheless, be obtained by constructing the bearing carriages in any of the modes usually practiced, provided that the fore and hind wheels of each of them be placed very near together."

I will remark, here, that it appears by the evidence, though not fully described, but only implied by the specification, that these bearing carriages, referred to as "bearing carriages in any of the modes usually practiced," were probably the bearing carriages of the eight-wheel wood cars, the eight-wheel trussel cars, the eight-wheel car Columbus, &c., invented and constructed by Conduce Gatch. That, however, will be adverted to in another part of the case. I will now confine myself simply to his theory, and the manner of connection or mechanism, which he describes and claims to carry that theory out.

At page 115—folio 17, he goes on to say: "I do not claim as my invention the running of cars or carriages upon eight wheels, this having been previously done; not, however, in the manner or for the purposes herein described, but merely with a view of distributing the weight, carried more evenly upon a rail or other roads, and for objects distinct in character from those which I have had in view, as hereinbefore set forth. Nor have the wheels, when thus increased in number, been so arranged and connected with each other, either by design or accident, as to accomplish this purpose. What I claim, therefore, as my invention, and for which I ask a patent, is the before described manner of arranging and connecting the eight wheels, which constitute the two bearing carriages with a railroad car, so as to accomplish the end proposed by the means set forth, or by any others which are analogous, and dependent upon the same principles."

The issue tried in the Troy case was, whether the Chapman car was, or was not, like the above theory and mechanism of Winans. Our issue is, whether the defendants' and Chapmans' are alike; and whether they are both different from Winans'.

We say that they are not like Winans', but are like each other.

The defendants in the Troy case said that Chapmans' was like Winans', and the plaintiff said it was not; and therefore the whole value of the evidence there, was lost in the improper manner of presenting the car, as we think.

It was lost to us, because it was not brought properly to bear upon the merits of the case; and it was lost to us, because it was misunderstood by the court, jury, and counsel.

Our position is, that the Chapman car is like the defendants', and not like the plaintiff's.

Judge NELSON.—You admit then that the Chapman car is not in evidence against the novelty of the plaintiff's invention? The ground taken in the Troy case was, that all the essential principles in the Winans' patent, were embodied in Chapmans' description, and therefore there was no novelty in it.

Mr. HUBBELL.—If this court takes the view that Winans' claim does include our car, then it must include Chapmans', which will destroy his patent; and our view is, that the defendants' is as much unlike Winans' as Chapmans' is, and the Chapmans' and defendants' are alike, and therefore they neither infringe. But it is not for us to presume that the court will construe the patent as we do. The issues are entirely different in the Troy case and this, and the mode of presenting the testimony, and the effect to be sought from it, is entirely different. In the Troy case, the effect sought was to destroy Winans' patent by means of the Chapman specification. Here, the effect sought is to enlighten the judicial mind, and protect the defendants' car by Chapmans', and show that Winans' patent should be construed on its own theory and mechanism, and if construed beyond that, it would be too broad, and void.

Judge NELSON.—Well, that is a fair argument.

Mr. HUBBELL.—It is not a patent which secures the embodiment of a principle; it simply secures a particular manner of connecting and arranging the wheels by the mechanism *described*, and which manner and mechanism the defendants do not employ.

The defendants' principle and physical doctrine is the same as stated in substance by Chapman, in his specification, and shown in his drawing, namely: That the axles of the wheels shall be held parallel to each other, in a rigid rectangular wheel-frame, consisting of two side-pieces and three cross-pieces; that the bearing-points of the wheels, on the rails, shall be equally distant from each other; that is, equal to the guage of the track. These axles, being thus held parallel, and the wheels square, the axles are always, as near as may be, parallel to the mean radial line between them; and the wheels, therefore, in line with the rails. That these transoms have each a centre-pivot, transom-plate or centre-bearing and side-bearings; and two of them are placed at a convenient distance apart, under the body, say about seven feet from th eends, laying the weight equally upon both. Hence they swivel to suit the curves, distribute the weight, and run safely and smoothly.

I will merely refer to the authorities on the Wood engine, which we find was presented in the Troy case, in the same manner as the Chapmans' car was. The Wood engine, described in Wood's Treatise, in 1825, is next in order of time, and it was printed and published to the world nine years before the date of the plaintiff's patent. The plaintiff produces John Elgar, with a version of this engine entirely different from the version in the Troy case, put forth by plaintiff. In that case, the trucks were admitted to swivel on a king-bolt, in the middle of the cross-piece; but his defence was, that the wheels were not as close together in each truck, and the trucks as far apart, as they may be; and the engine was there alleged to have the extraordinary capacity of lifting itself up off the track into the air, before the wheels would slip to consume the power by velocity. The success with which these nonsensical theories and doctrines were advanced in that defence, by plaintiff, as to that engine, is really wonderful. Science stands aghast at the spectacle, and the best of practical engineers are confounded at such pretensions. Mr. Elgar, the plaintiff's bosom friend, is ready to expound every new doctrine, starts another view of the construction, and says that when Wood describes a bolt in the middle of the crosspiece, to allow the frame and wheels to conform to curvatures, he intended to describe two bolts, outside, one at each end of the cross-piece, with a transverse motion, like their model, which Elgar says is a lateral motion, and which would inevitably prevent the engine from running on any part of a curve. Mr. Elgar is contradicted in this, by the plaintiff's own witnesses, in the Troy case, viz., by Mr. Hibbard, Troy case, p. 88, and Peter H. Dryer, Troy case, p. 89. Both these witnesses contradict Mr. Elgar. Mr. Hibbard, plaintiff's witness, p. 88 defendants' print of Troy case, and p. 94 in Appendix A, to plaintiff's evidence, says: "I find in the treatise and drawing the two trucks are connected with the frame or body by *bolsters* or *king-bolts*, as in the model. By divesting the carriage described and represented in Wood's Treatise of the cog-wheels and apparatus, the obstacle to the free motion of the trucks would be removed." There, one of the plaintiff's witnesses testifies that it has *king-bolts*, which is a contradiction of Mr. Elgar's evidence. Peter H. Dryer, page 89 defendants' print of Troy case, and p. 96 of plaintiff's Troy case, in Appendix *A*, says: "I understand the description and drawing in Wood's Treatise. I do not consider that a practical locomotive carriage could be made from the drawing; there are two trucks, the inner axles being connected by a cog-wheel; the motive power connected with the inner wheel; the free motion of the trucks would be impeded by the meshing of the cogs. There would be considerable strain on the *king-bolts*." "On the king-bolts!" Now, Mr. Elgar says that it has no king-bolt. He goes on to say: "The amount of the strain would be the whole leverage from the *king-bolt* to the cog-wheel." This witness, therefore, also contradicts Mr. Elgar.

I will now refer to the defendants' witnesses, to prove the principle and essential arrangement and construction of the defendants' car to be embodied in this Wood's engine of 1825.

John Edgar Thompson, defendants' No. 3, pages 1, 2—folios 4, 5.

George W. Smith, defendants' No. 3, pages 73, 76—folio 292; page 75—folios 298, 303.

Wm. J. McAlpine, defendants' No. 3, page 94—folios 375, 376.

Walter McQueen, defendants' No. 3, page 101—folio 402.

George S. Griggs, defendants' No. 3, page 106—folio 424.

All these witnesses contradict Mr. Elgar.

Mr. Hibbard and Mr. Dryer, two of the plaintiff's witnesses, as well as the witnesses on the part of the defendants, put exactly the same construction on the Wood engine that we do, as regards the *king-bolt;* and it was not presented, in the Troy case, in the same light in which we think it should have been presented. We are of opinion that it should have been presented in the same view as we present it here, and as we have presented Chapmans' car here; subject, however, to the same effect of invalidating the patent, if the court put a broader construction on the plaintiff's claim than we do.

TREDGOLD CAR.

The next defence to which I refer is the printed publication:

"*A Practical Treatise on Railroads and Carriages, &c., &c. By Thomas Tredgold, Civil Engineer, Member of the Institution of Civil Engineers, &c. Printed in London,* 1825,"

This is the next work in point of time on which we rely, and particularly the following parts of the publication:

[Page 93 of Tredgold.] "Carriages for common railroads are made strong, to resist the shocks they are exposed to at every change of velocity; and it is necessary to make the *parts,* which come in *contact*, solid pieces, extending a little more than the length of the body of the carriage, and hooped at the extremities, to prevent splitting; but carriages for *passengers*, and for various kinds of goods, must be provided with *springs*, to reduce the force of these shocks."

[Page 94.] "Small carriages must obviously be both heavier and more expensive, in proportion, than large ones. But as the stress on a *wheel* must be limited on a railroad, we cannot much enlarge the carriages without *adding* to the number of *wheels*."

Eight-wheeled Carriages.

"When a carriage has more than four wheels, the *body* must be *sustained*, so that *its pressure* may be *divided equally* among the WHEELS. In the case where eight wheels are applied to support *one* body, if the body rests upon the *wheel-frame* of each set of *four* wheels, in the middle of its length (see fig. 26, plate iv), and is connected with those frames so as to allow the greatest possible change of level on the rails, it is obvious that each wheel must bear an equal pressure."

[Page 95.] "The load on each *wheel* must be limited to suit the strengh of the rails; it will seldom exceed two *tons* on a *wheel*, nor be less than half a ton. The size of the axles may therefore vary from twenty-two inches to thirty-five inches. Perhaps the most advantageous load will be about one and a quarter *tons* on each *wheel*, which will require an axis of three inches diameter."

[Page 179.] "Plate iv. fig. 26, a diagram to show how a wagon may be made with eight wheels, so that the stress of each *wheel* on the rails of a railroad may be equal. The body of the wagon rests on the wheel-frames at A A, and is connected to them by an axis, on which the frames *turn*, when, from any inequality, the axis of the wheels are not in the same plane. [See page 94."]

[Pages 12, 13.] "'The Helton railway is one of the principal ones." In some parts near the Staiths we observed malleable iron rails, in fifteen feet lengths, supported at every *three* feet. (See figs. 9 and 10.) They are three and a half inches deep, in the middle, between the supports, and two and a quarter inches in breadth, at the upper surface; one yard in length weighs about 28 lbs.

"The wheels of the coal wagons are two feet eleven inches in diameter, with ten spokes, and weigh two and three-quarters cwt.; and their *axles* are three inches in diameter, and *revolve* in *fixed bushes*."

"The weight of the engine is about eight tons. (See fig. 2, plate 1.)"

"The boiler is supported on the carriage by four *floating* pistons, which answer the purpose of *springs* in equalizing the pressure on the *wheels*, and softening the jerks of the carriage."

[Pages 118, 119, 120, 121.] "The extent of land required for a railroad must depend on the breadth and number of the tracks. The breadth of the track has rather been determined by opinion, than as a question arising out of the circumstances of the case. But it must be obvious that the breadth of the track ought to have some relation to the height of the load, in order that the carriage may be always in stable equilibrium on the rails; and in railroads there is another circumstance to be considered—the pressure on the rails should not be materially altered by any slight depression of one side of the road. It may be taken as a general rule, for the width between the rails for carriages traveling at a greater speed than five miles per hour, that the centre of gravity should not be higher in proportion to the breadth between the rails than as 1 is to $1\frac{1}{2}$, but they are often so constructed as to be equal, or nearly equal, to the breadth between the rails; and with this proportion, in slow motions, no ill consequence may probably occur; but in *rapid* motions, the centre of gravity must be kept at least within the limits we have mentioned, or there will be much risk of the carriage being overthrown by a very small obstruction. On a common road, the great resistance at the surface of the wheels, and the force of the moving power, tend to keep the carriage from upsetting; but on a railway, the smallness of the moving force, and little increase of resistance to the wheel which takes the stress, render them insufficient to balance the momentum the carriage acquires by striking on an obstruction; besides, the connection of the moving force is not so favorable for drawing the carriage back to its position. All these circumstances demand the serious attention of the engineer who has to conduct a railway, where the carriages are to proceed at the rate of ten miles an hour. The width between the rails being, therefore, dependent on the height of the centre of gravity of the loaded carriages, and this again varying with the nature of the load and the velocity, it will be obvious we cannot do better than make the breadth between the rails such that, by disposal of the load, the centre of gravity may be kept within the proper limit in either species of vehicle, whether swift or slow. And it would be desirable that the same breadth and the same stress on a wheel should be adopted in all railways. We would propose four feet six inches between the rails for heavy goods, and six feet for light carriages, to go at greater speed." "In all railroads, it is necessary that there should be passing places at certain parts of the road, and in single tracks they should be very frequent."

[Page 127.] Example.—"If the distance of the supports be three feet, &c."

[Page 130.] "If the distance of the supports be three feet, &c."

[Page 126.] "*The distance between the wheels of the carriages should be such that the unsupported part of a rail should have to carry only one wheel.*"

[Page 101.] "In proportioning the body of a carriage, it should be kept in view that the load should be as low as possible."

[Page 173.] "For *carriages on springs*, and steam carriages, consider the stress, &c."

[Page 174.] "Where the heaviest *carriages are on springs*, or floating pistons, two-thirds of the actual load on each wheel may be considered the stress on the rails for ordinary purposes, &c."

[Page 12.] "The Helton railway is one of the principal ones." "The rails of the straight parts of the railway weigh 61 lbs. each! but this was found to be too slight for the *curved* parts; therefore, the strength of the latter has been increased, &c."

[Page 135.] "When a considerable degree of *curvature* is given to a railroad, the rails of the outer curve should have a slight *rise* to the middle of the curve; and the rails should be stronger in a lateral direction in both lines. The object of making a slight ascent to the middle of the curve of the outer rail, is to counteract the tendency of the carriage to proceed in a straight direction, without its rubbing so forcibly against the guides, as we have observed in cases where roads have had a considerable curvature."

[Pages 42, 43.] "It often happens that a great part of the resistance of the rails arises from the lateral rubbing of the guides of the wheels; therefore it is desirable to give the wheels a tendency to keep in their path, with as little assistance from the guides as possible. For edge-rail carriages, this may be accomplished by making the rims of the wheels slightly conical, or rather curved, as shown in fig. 24. The carriage will then return of itself to its proper position on the rails, if it be disturbed from it by any irregularity."

The car shown in the diagram, fig. 26, is about eighteen feet in length; and hence the wheels, under the rules laid down, are correct, as there shown, at an equal distance apart on the rails.

If the car be made longer, or a longer body be placed on the same trucks, the same rules, to attain the same expressly required conditions and results, will cause the wheels to remain the same distance apart in the trucks, and the trucks to be placed under the ends of the body, the same distance from the ends; and, *consequently*, the *trucks* to be *further* apart, proportionate to the increased length of the body, such as demonstrated by the long body, A 2.

Tredgold *cannot* be considered as intending always, or with longer car bodies, to place the wheels at an equal distance apart. He expressly states his object to be, to divide the pressure *equally* among the *wheels*, so that it shall be about two *tons*, or, perhaps, most advantageous, "about 1¼ *tons* on each *wheel*." He requires both the body of the car and the load to be *low*, so that "the centre of gravity should not be higher, in proportion to the breadth between the rails, than as 1 is to 1½." [Page 118.]

On a four feet eight inch track, this would make the height to the centre of the load about 3 feet. The rules laid down, and the ordinary height of wheel in use at the time, show the diagram to be in the proportion of a car about 18 feet long; and to preserve the same application of the rules, laid down in a longer car, would require the same trucks equally near the ends of the longer body, and having the, consequently, greater distance between the trucks alone.

To suppose the diagram to represent a car 60 feet long, would make the body about 15 feet high, and the wheels 8 or 9 feet in diameter, on a 4 or 8 or 6 foot track, which would be ridiculous, and violate the rules of construction, as to height of load, weight and arrangement of wheels, which Tredgold lays down. He NEVER intended to place the wheels at an equal distance apart on the rails, in a car as long as 40 or 60 feet. He shows a *practical* car, 18 *feet* long, and the *same trucks*, are to be used with *longer* bodies, when required. The rules he lays down, as to the height of the body and load, the position of the wheels on the spans of the rail, the division of the pressure in *tons* among the *wheels*, to relieve the rails of too much stress on a wheel, the proportions of his diagram, &c., all show this.

Also, a car body, 40 or 60 feet long, with trucks holding wheels at an equal distance apart, would have the trucks so *heavy* and *large* as to be in direct violation of his purpose, of having but little stress from each wheel on the rail, and of running on a railroad, as generally built. His rules and instructions, laid down, are the practical rules now followed in building and using the eight-wheel car; and they are not changed, whether the car be 18, 20, 40 or 50 feet long. His diagram shows a car about 18 feet long. Putting a longer body on the same trucks, under the same rules, does not alter the principle, essential arrangement and construction, nor the operation. The same eight-wheel car is now used, from 15 to 50 feet in length, for carrying different articles, or objects, animate and inanimate, such as wood, water, freight, live stock and persons.

In consequence of intimations heretofore thrown out by the court, I will further refer to the Tredgold car, very briefly, in seven propositions.

It is fully proved on behalf of the defence, and model A is the representation of it; it is proved by thirty-seven of the defendants' witnesses as the car described in Tredgold's Treatise, published in London in 1825, a period of *nine* years before the date of the plaintiff's patent.

The court will understand that we present Tredgold in the same manner that Chapman has been presented; subject to the construction placed upon the plaintiff's patent by the court. The propositions are:

1st. Tredgold describes an axis to each wheel-frame, and two of these wheel-frames of four wheels each.

2d. The car he describes is to run on a railroad, and is represented in sketch as a known principle.

3d. *Query.* What was a wheel-frame of four wheels, with an axis for a railroad car, before the time that Tredgold wrote—1825?

4th. In 1814, Chapman describes and shows it as consisting of two side-pieces, and three cross-pieces; central and also side-bearings, and a vertical pivot, or king-bolt, or axis, as it is called.

5th. This, then, is the same wheel-frame, of four wheels, described by Tredgold, with a central bearing *alone*—as he shows—which divides the pressure equally among the wheels, and conforms to any change of level,—both on the straight track and the curves—one set of wheels being shown under each end of the car.

WILLIAM W. HUBBELL. 103

6th. The railroads of that day, 1825, were more curved than the railroads of the present day; as shown by the testimony of John Edgar Thompson and Geo. W. Smith, and expressed by Tredgold himself, on pages 135, 158.

7th. It is drawn by a chain coupling, from the middle of the end of the body; the same as shown in the drawing, plate 1, figure 1, in Tredgold,—that being the only mode shown to draw cars in trains.

These propositions cover and show what the car is, precisely.

I will now refer to the witnesses on this point, and give the pages and folios that contain their evidence:

Names.	Defts' No.	Page.	Folios.
1. Wm. B. Aitken,	2,	5,	17, 18, 19
2. Stephen Ustick,	2,	6,	21, 22
3. Oliver Byrne,	2,	10,	37 to 45
4. Albert S. Adams,	3,	115, 116,	460, 461
5. John Edgar Thompson,	3,	2, 3,	5 to 11
6. Wm. Pettit,	3,	13, 14,	49 to 55
7. Asa Whitney,	3,	14, 15,	56 to 59
8. John Murphy,	3,	10, 11,	38 to 44
9. Jacob Schryack,	3,	40,	159, 160
10. Henry Waterman,	3,	66, 67,	263 to 267
11. Wm. J. Ragland,	3,	69,	273 to 276
12. Geo. Stark,	3,	70 to 72,	281 to 286
13. Chas. B. Stuart,	3,	90, 91,	358 to 361
14. Geo. W. Smith,	3,	74,	296 to 299
15. Vincent Blackburn,	3,	97, 98,	388 to 391
16. Wm. J. McAlpine,	3,	94,	376 to 378
17. Geo. S. Griggs,	3,	107, 108,	425 to 429
18. Walter McQueen,	3,	101,	403 to 407
19. James H. Anderson,	3,	119,	473 to 475
20. Albert S. Adams,	3,	115,	460 to 463
21. Asahel Durgan,	3,	127,	506, 507
22. Jno. B. Winslow,	3,	122,	485 to 487
23. Hy. W. Farley,	3,	130,	519, 520

Names.	Defts' No.	Page.	Folios.
24. Jno. Crombie,	3,	127,	506 to 508
25. Wm. C. Young,	3,	150,	600, 601
26. Wm. P. Parrott,	3,	155,	618 to 623
27. Chas. Minot,	3,	160,	641 to 644
28. Godfrey B. King,	4,	80, 81,	302 to 306
29. Conduce Gatch,	2,	22, 23,	88 to 92
30. Jno. C. A. Smith,	2,	25,	99 to 102
31. Isaac Knight,	2,	27, 28,	111 to 113
32. Geo. Law,	2,	29, 30,	120 to 122
33. Edward Martin,	2,	35,	145 to 149
34. Geo. Beach,	2,	38,	157 to 160
35. Stephen W. Worden,	2,	39,	163, 164
36. L. R. Sargent,	2,	40, 41,	169 to 172
37. Jno. Wilkinson,	2,	43 to 45,	181 to 194

These thirty-seven witnesses embrace the three classes of the best witnesses, and the best practical men in the country. The three classes are:

1st. Men of education and experts, who understand the principles of science, in their abstract character and mechanics.

2d. Superintendents who have a controlling direction over machinery, to say what shall, and what shall not, go on a road, and how it must be constructed.

3d. Men who perform manual labor and make the machinery, and know exactly how it should be constructed, and the reasons why constructed in the particular mode, and are called practical mechanics.

I have now a few remarks to offer in relation to the plaintiff's witnesses. To these 37 practical and experienced men, the plaintiff opposes 4 witnesses: William C. Hibbard, Jeremiah Myers, Lee McComas and John Elgar.

William C. Hibbard, it appears, is a patent agent, and professional expert. He does not state that he ever located or constructed a railroad; he does not explain, or state, that he knows the nature of the changes of level in a railroad; nor does he say that he is acquainted with the nature of railroads existing in England in 1825. He speaks of curves of short radius, at folio 363, of the plaintiff's testimony, as being a new condition, which, with more speed, indicated wants in engineering that did not before exist, and to meet which, is the object of the plaintiff's invention.

Whereas, we prove that the roads constructed before 1825 were more curved than the Baltimore and Ohio road, and roads in this country; and, therefore, the roads of that day, necessarily required the eight-wheel car to have two swivelling trucks, such as Chapman and Tredgold describe, and as are now in general use; and that the speed is due to the locomotives, as improved in power—not to the cars. He assumes that the doctrines set out in Winans' specification are correct, and forms his opinions on such assumption. He knows nothing practically about railroad machinery; and we prove by our experienced witnesses, that the doctrines, and also the mechanism, of the specification are unsound and impracticable; particularly on curves, and at high speed.

He talks, at folio 369, of the "*momentum* of the *shocks*," and then gives a computation of their increase of intensity with an increase of velocity of the car, when, in fact, a shock has *no momentum;* and the velocity of the car being increased, tends to carry it over the joints of the rails, without falling or settling into the spaces between them, and, hence, diminishes the shock.

A shock *has a measure of force*, which is determined by multiplying the weight by the velocity in feet per minute; but the shock itself has *no momentum*.

This Mr. Hibbard, at folio 371, then says: "These difficulties are remedied by constructing the car in the manner fully described in Winans' specification;" and then he follows the substance of the specification, as to placing the wheels close together, and the bearing carriages at a remote distance from each other; and adds, among them, the application of the tractive power to the body alone, as vitally essential, in his judgment, to the proper construction of an eight-wheel car; while the fact is, that the specification does not describe any mode whatever of drawing the car; and really intelligent engineers, who do know how the eight-wheel car is and should be constructed, describe the same, both in its construction and action, and demonstrate it to be different from the specification of Winans. At folio 385, he states that those conditions of speed of transit, or curvature of road, have not come to his knowledge from the books, or any other source—by which the invention of the plaintiff was required—the truth being, that the Tredgold book itself, is indexed "*curved roads*," pages 135 and 138 ; and the pages referred to, both contain descriptions of curves and allusions to increase of speed ; and all the roads in England, at the time, were very much curved, as proved by Mr. Thompson and Mr. Smith, and steam engines of different degrees of power and speed in use on some of them, thus making his ignorance and want of discrimination that improvements in locomotives gave the increase of power to draw the same cars at higher speed, and his carlessness in not examining the book, a basis for an opinion in a matter of such vital importance as this. At folio 386, he gives his opinion, or the appearance to his mind, as he states, that the invention of the plaintiff was for the purpose of remedying certain defects in the movements of cars, while those before existing were for the purpose of remedying a *defect* in the strength of the *rails*, and that *this* is a broad *distinction* between the *invention* of the plaintiff, and those before existing.

And Mr. Hayes, also the plaintiff's witness, swears that the wheels are further apart, in the trucks on the Baltimore and Ohio road, than stated by Winans' specifications, for the purpose of *distributing the weight* of the bearing cars *on the rails*.

Thus, one of the plaintiff's witnesses, giving evidence, upon practice, contradicts the other, giving evidence upon the theory of the specification.

I must remark, further, that with regard to the latter part of Mr. Hibbard's opinion, it would be the work of a life-time, left unfinished, to find out how an *invention* in a *car* can consist in the *purpose* of *remedying* a *defect* in the strength of the *rails*. The invention, I believe, is generally, and so far as I know, considered as continued in the mechanism of the car, to effect the purpose by means of the operation or principle; and the judgment of the effects of inexperience, or manifest want of common knowledge of the subject, such as Mr. Hibbard's, founded on a limited conception as to purpose,—and assuming the doctrines of the specification to be true, without actual experience and knowledge—certainly is not expert evidence, on which a court can rely, as against thirty-seven practical, sensible men, on the same subject. Mr. Hibbard's remarks, in his testimony, on the vital importance of drawing the car by the body, seem to be founded on a misapprehension, on his part, that Mr. Winans' patent secures to him this mode of drawing a body with trucks under it; while the fact is, as appears by the specification itself, and from the testimony, that no mode whatever of drawing or moving the car is described or claimed by Mr. Winans.

With reference to Mr. Myers, the plaintiff's witness, *he* states that cars could be constructed from the plaintiff's patent, without stating the construction, or conditions, or principle, in terms. It is therefore impossible to determine what is in the mind of the witness, as to construction, arrangement or principle. He is one of the class of experts who offers to decide a matter, without giving the basis of reasons and considerations upon which his conclusions depend; and hence, his declaration cannot be examined and weighed by the court. He gives a profuse sprinkling of scattered and incoherent extracts from different parts of Tredgold's book, taken from entirely different headings, on different subjects—and finally takes up half of the description of Tredgold on the eight-wheeled car—that part on page 94, omitting that on page 179—and declares that he cannot find stated, in said treatise, any object or purpose for which the plan of the car described in said letters patent would be particularly useful. This shows at once that he perverts the true construction of the Tredgold car, as described, from its legitimate purpose. While in fact it is expressly stated on page 179—the part which he has omitted, taken in connection with page 94, and the diagram, as to run on a railroad, &c. Such a witness as this, as an expert, who avoids and perverts the examinations of the article, subject "eight-wheel carriages," in Tredgold, can have no weight, as against practical men. He is contradicted in his statement that eight-wheel cars, constructed and arranged after the plan of a Baltimore car, at Philadelphia, and as directed in said letters patent, were placed on the Boston and Providence railroad.

The Philadelphia cars are proved to have originated at Philadelphia, and to have been successful from the commencement; and the officers of the Boston and Providence road—Mr. Lee and Mr. Griggs—prove that no such car, as described in Winans' specification, was ever put upon the road.

So much for Mr. Myers. He pretends to examine Tredgold, as to eight-wheel cars; and, after hopping about on to six or eight distinct and different subjects, finally ends by finding half of the description on the subject, and then states, or implies, that no such purpose is there expressed as the ordinary purpose of an eight-wheel car to run on a railroad. He says that he built cars like Winans', and does *not* describe the *machinery;* and is *contradicted* by the engineer of the Boston and Providence road as to his statement concerning cars on that road.

In the investigation of a question like this, it is the *machinery* that is to be examined. The *machinery* is the subject of a patent; *not* the *purpose,* and *not* the *principle,* incidental to the machinery; and where a witness like this omits to give the construction of the machinery altogether, he is not entitled to any consideration whatever from this court on the subject of this suit.

Lee McComas is the next expert of the plaintiff. He says that his business is that of a pattern-maker and machinist; that he has been employed as such for the last 23 years; but by whom, or where he has been employed, is concealed.

Is he not Ross Winans' pattern-maker?

Mr. LATROBE.—We admit that fact—that he is.

Mr. HUBBELL.—Then it should have been stated in the affidavit, and not studiously omitted, as there appears. He makes up his own specification, at folio 323, in these words: "*that is* spread over the eight wheels, each wheel bearing its share of the load, *provided* the *load* be properly divided;" and, from this, makes the plaintiff's black model.

No such condition or purpose is stated in the Tredgold specification, as to the *eight*-wheel car. It is only stated as to the six-wheel car; and hence, this model is made for a purpose, and from a specification, not contained in the book. It is an intentional perversion, and entitled to no reliance whatever.

The next witness is Mr. Elgar. Neither does he state how he is employed. Is he not in the employ or pay of the plaintiff?

Mr. LATROBE.—No. He is an engineer, and surveys gentlemen's country seats to put down water-rams. He is a surveyor.

Mr. HUBBELL.—He asserts that car builders cannot do, from the whole book, what they themselves testify they can from pages 94 and 179, and the other references, and have done, as is directed there.

He describes no "features and advantages," but leaves them undescribed, and hence we do not know what Mr. Elgar means.

He testifies to a model marked in a certain manner, which is not produced so marked, and gives us his conclusions, without the mechanism on which they are founded.

It is impossible, therefore, to give any weight to his opinion; for the means or channel of mechanism, and the certain assumed purposes and conditions upon which it was based, do not appear for investigation.

Any man, however ignorant or mistaken, or biased on this particular subject under investigation, could swear away another's rights, if such testimony as this were entitled to weight in a court of justice.

110   ARGUMENT OF

These are all the plaintiff's experts, and we respectfully submit that they are entitled to no weight whatever, and that we have in the case the *Tredgold car* (as well as the Quincy car, the Chapman car, and the Allen car) proved conclusively to be substantially the same in construction and operation, or principle, as the eight-wheel cars built by the defendants, before the date of the plaintiff's patent, and before 1830.

QUINCY CAR.

The Quincy car is admitted by the plaintiff's witnesses to have been in use in 1830. John L. Cofran misdescribes the bearing of the bolsters, as being upon and across the truck, whereas, in fact, it had side-bearings, as shown in the model; and it is proved positively by the defendants' witnesses, who made and used it, as represented by the model before the court, to have been constructed and run on the Quincy railroad, by Gridley Bryant, in the summer of 1829, more than five years before the date of the plaintiff's patent; and that it and similar cars have been openly and publicly used there, continually, until the present time.

This is proved by four witnesses.

There are also seven witnesses, as experts, who prove that the principle, construction and arrangement of the eight-wheel car, built by the defendants, and in general use, and this Quincy car, are essentially the same.

The plaintiff's experts show that they are incompetent, on the subject of railroad structure and railroad machinery. And the rights and interests of these defendants, and of railroads, against which suits are brought, which will be influenced by a decision in this case, we submit, are not to be sacrificed in the courts of the United States by such incompetent, impractical men, possessing and showing not the slightest practical knowledge of the conditions, nature and principles of railroad machinery.

The one of them a professional expert, or affidavit speculator. An ther of them a surveyor of water-rams, and an old acquaintance and bosom friend of the plaintiff, imbued with all the feelings of freindship and all the notions that contact with the plaintiff may have impressed upon him.

Another, the pattern-maker of the plaintiff, and dependent upon him for his daily bread ; and the other, an unsuccessful car builder, whose ideas are as incoherent and crude as the human mind could be on such a subject, possessing neither order of thought, knowledge of fundamental principles, nor capacity to describe, nor apparent willingness to tell the whole truth, and who, as we shall show in another branch, has stated that which he ought to know is erroneous, and concealed that which he ought to have stated.

If such men, inexperienced as they are, and circumstanced as they are shown to be, are to be relied upon, as against sound, experienced, practical men, trials in courts of justice become a mockery, and the vested rights of the public at large are mere fleeting shadows.

PLAINTIFF'S WITNESSES.

The next point to which I will advert is, the weight to be attached to the plaintiff's witnesses; and here, I am sorry to say, my duty becomes a painful one. Such a tissue of errors and contradictions as is contained in the evidence of the plaintiff's witnesses, I never before remember to have met with. They not only severally contradict themselves, but they contradict each other in different directions on the same subject; stating things from the plaintiff himself, throughout, which ought to have been examined, in order to have ascertained their truth or falsity, before affidavits were made in relation to them. I propose to examine the testimony of each one separately, and to show that things have been stated to be true, which ought to have been known were erroneous; that at one time, one thing has been sworn to, and at another time, something else contradictory ; that they have testified that they have known, or believed things lately, which, some years ago, they testified that they did not know ; and to show, also, that they have contradicted each other, as to material matters in this case.

The first to which I will refer, in this connection, is the affidavit of Ross Winans himself. On page 3, second paragraph, of plaintiff's proofs, he states: "And this deponent further says, that *on obtaining the letters patent* for said invention, *he took measures to make it publicly known, and to introduce his improvement into general use.* He *published the specification in the American Railroad Journal, a paper published in New-York city, having an extensive circulation among all parties interested in the existing and projected railroads of the country.*"

This affidavit was before Judge Conkling, at the last hearing, and the judge thought that a very wise measure had been thus taken by Mr. Winans; and, at that time, we did not know that it was unfounded and erroneous. Much reliance was there placed upon it by the plaintiff, and great weight by the court. The motion failed, however, notwithstanding.

For the purpose of investigating the circumstances, with a view to ascertain whether such a publication had been made in the American Railroad Journal, Mr. Randall, as his affidavit will show, very fully, and with much time and care, searched the publication, with the results which his evidence will demonstrate. He says: "that he is 43 years of age, and resides in Syracuse; that he is by trade and profession a printer and editor, and is accustomed to the examination of files of newspapers and printed publications. I have examined the files of the American Railroad Journal, referred to by Ross Winans, in his affidavit in this case, for the purpose of finding the publication of the specification of his patent of October 1st, 1834, on which this suit is brought, and which he testifies was published in the said American Railroad Journal, published in New-York city, on obtaining the said letters patent. I very fully and carefully searched the said Railroad Journal, in the library of the Franklin Institute of the State of Pennsylvania, in the Mercantile Library of the city of New-York, and in the office of publication of said Journal, in New-York city, for all the years from 1832 to 1847, inclusive, and being through a period of fifteen years, for the purpose of finding said specification published therein, but did not find the same, nor any allusion to the invention of an eight-wheeled car by Ross Winans; and if the said specification, or any publication of Ross Winans, similar to it, had been published there, I believe that I would have found it. Not having found it, after the search I made, I am confident that Ross Winans is mistaken in his statement, and that no publication of his specification of October 1st, 1834, was made in the said American Railroad Journal. The only specification in the said Journal, of a patent issued to Ross Winans, is of an improvement in wheels for cars and locomotives, to be used on railroads; the date of the patent being 1833, and which is published in the volume of 1834—this being the specification of another patent entirely.

"I also searched for the publication of notice of the application of said Ross Winans, for an extension of said patent of October 1st, 1834, in the volume or file of the Sun newspaper, published in New-York city, in 1848, and that of the National Intelligencer, published in the city of Washington, in 1848, in the respective offices of said papers, and, after a very full and careful search, did not find such notice published in either of said papers; and not having found the same, I am confident that the said notice was not published in either of them. I also got the owners of said papers to examine their books, to see if any charges had been made on their respective books for said publication, but they could find no charge, and expressed their belief that no notice of the kind had ever been published in their respective papers. I have no interest in this suit."

Notice was given to the solicitors on the other side, at the time that this affidavit was served, to produce that American Railroad Journal, in which the specification was alleged to have been published.

The plaintiff's counsel now, when they find that we have proved the assertion of Winans erroneous, *say* that said specification was not published in the work in which Winans swore that it was; but, as an excuse, say that it was published in the Franklin Institute Journal, in Philadelphia, and a volume of that work is produced for some other purpose, I believe to show a publication concerning the Helton railway; but there has been *no volume* of the Franklin Institute Journal produced to show that their present statement is true, or that Mr. Winans had even negligently and unintentionally committed an error in his statement.

Mr. Randall's affidavit proves conclusively that no such publication was ever made in the American Railroad Journal, and they now admit the fact—and therefore Judge Conkling was misled.

Such an erroneous statement as that is inexcusable, to my mind. Mr. Winans should have satisfied himself as to the fact, by a personal examination of the Railroad Journal, before swearing to such a statement. He should not have been so negligent and careless as to make an assertion like that, without first ascertaining that it was correct.

Mr. Winans further says, that " he procured its adoption on the Baltimore and Ohio railroad, where it had been first tried, improved and perfected by him; he then endeavored to introduce it on the roads leading from Boston, and one of these, the Boston and Lowell, in 1836, took and paid him for the license under which they have since been operating it." This also is erroneous; for, in February, 1834, it appears that a general assignment of all the rights he had, or might have, was made to the Baltimore and Ohio Railroad Company, and there was no express purchase or mention of this patent. It had no existence at that time. And we have shown, by the testimony of Mr. Schryack, the foreman of the car establishment of that road, that the car described by the specification was not adopted on that road.

Also, the same is shown by the testimony of Hy. Shultz, defendants' No. 4, folios 156, 157, 158.

Eight-wheel cars, of some kind, we know, were used on that road; but, the question is, whether they were the same as described in Winans' specification; and the testimony shows that they were not.

Mr. Schryack's testimony further shows that Winans' specified car was subsequently tried with cast-iron saddles on the springs; and even with that addition and assistance, it failed.

The statement that the "Boston and Lowell road have been operating it," is disproved by Waldo Higginson, the engineer of that road, who testifies that such a car has never been on the road as the one described in Winans' specification; and that four patents were purchased of him, collectively, for a thousand dollars.

Therefore, as no such car as that described in Winans' specification was ever made for or used on the Boston and Lowell road, the consequence is, that the company has *never* been "operating it under the patent;" and the statement of Winans, in this behalf also, is both equivocal and erroneous.

Mr. Higginson is also confirmed in his statement by Wm. Raymond Lee, superintendent and chief engineer of the Boston and Providence road, defendants' No. 3, page 112—folio 147.

Mr. Winans states further, in his affidavit: "And the chief engineer of another, the Boston and Providence, ordered at your orator's instance cars upon his plan, which were afterwards countermanded from want of confidence in their success, although this company subsequently adopted, and have since been using them."

Mr. Lee directly contradicts such a statement, in the last reference to page and folio, where he says, that "*No such order as stated by Ross Winans was ever given at his instance for cars upon his plan, nor for any other plan.*"

On page 4, folio 4, Winans further adds: "About the same time he furnished, for the same purpose, several of his improved cars" (I infer that the intention here, is to impress the court with the opinion that these cars were constructed in the manner described by the specification) "for the Philadelphia, Wilmington and Baltimore Railroad Company, which said last mentioned company purchased and paid him therefor."

116   ARGUMENT OF

The testimony of Henry Shultz, defendants' No. 4, folios 64 to 68—pages 43, 44, proves that no such cars, as described in Winans' specification, have ever been on that road.

At folio 165, he says: " In the year 1837, I engaged in car building with the Philadelphia, Wilmington and Baltimore Railroad Company, at Baltimore, and have had charge of the building and repairing here ever since. I have built nearly all of the eight-wheel passenger cars that the Philadelphia, Wilmington and Baltimore Railroad Company use, this side of the Susquehanna river; and I understand the principles of construction and action of the eight-wheeled cars in general use.

" The eight-wheel passenger cars in use have swinging bolsters and draw-springs, the trucks are rigid rectangular wheel-frames, made as rigid and stiff as possible, and the cross bolsters and end pieces are secured by joint bolts. The distance of the body bolster from the end of the body brace is about seven feet, and the distance from the end of the box part about four and a half feet, and from thirty-five to forty-four feet in length. The distance between the centres or bearing points of the wheels in each truck on the rails, is made as far apart as we can get them, in a truck from seven and a half to eight feet long; they are further apart than the wheels of the Columbus were, and to get them far apart is an advantage—it makes the car run easier and steadier than it would if they were closer together. The trucks have male and female transom-plates, to connect them to the body and allow them to swivel, with a body bolt passing through them to hold them together. The male and female transom-plates, for the trucks to swivel, are in universal use on the eight-wheel cars on railroads, and with swinging bolsters on the passenger cars; I have not seen any without them. I have examined and understand Ross Winans' specification for an eight-wheel car, dated Oct. 1, 1834; at least I understand what Ross Winans means to describe by it, but it will not answer for an eight-wheel car; a single pair of wheels, as he states, would turn right around into the middle of the track; and putting the wheels close together, as he describes, would tend to do the same thing; the springs that he describes for coupling the wheels are good for nothing; they have been tried on the Baltimore and Ohio road, and condemned; they will not hold the wheels steady, and if one should break, it would destroy the whole truck; coupling the truck with the body by a king-bolt, like a common wagon, which he describes, is not the way in which eight-wheeled cars are built. They have male and female transom-plates, and the body-bolt may be taken out, and yet the car will run; the body-bolt is a security only in the eight-wheel cars; the swivelling is in the transom-plates. There is no attempt in Mr. Winans' specification to describe any mode of drawing or moving the car. The principles or conditions and construction there set out in that specification are unsafe, liable to accidents, and would not be suitable for railroad companies to use. I never saw his specification before, and have no interest whatever in the subject of this suit." Showing, again, that Winans' language is both equivocal and erroneous.

At folio 5, Winans continues: "Deponent says, however, that the improvement gradually came into favor; and the Newcastle and Frenchtown Turnpike and Railroad Company adopting and refusing to pay for it, a suit was commenced against them in the Circuit Court of the United States, for the Maryland district, in 1838—was most thoroughly contested before Chief Justice Taney, and although not resulting in a verdict, the jury failing to agree, it had the effect of inducing the defendants to pay to deponent the sum which he had demanded before the action was commenced."

Now, the testimony of James B. Dorsey shows, in defendants' No. 4, folio 292—page 77, that he planned and built the cars for the Newcastle and Frenchtown road, and that "they had rigid wheel-frames for the trucks, well ironed, to make them stiff;" which is essentially different from the car described in Winans' specification.

In regard to the ruling of Chief Justice Taney being in his favor, as he states, we have examined it, and we differ on the subject; and although it is a matter of opinion, however, it is set up here with as much assurance as though it were expressly and directly on his side; whereas, as we understand that opinion, it is *adverse* to the plaintiff.

He then states, at folio 9, page 5, in relation to the Baltimore and Susquehanna Railroad Company, that "small, however, as was the sum claimed by the deponent, the amount offered by them was still less, and as they fully admitted the value of his improvement." Mr. McGraw and Mr. Hollins' testimony show that that company did not admit that it was an improvement. They offered a small sum, or to docket a suit *at home*, where the plaintiff, the witnesses, and they all were; but no! Mr. Winans flees from the justice of the district of Maryland; he flees from Chief Justice Taney's instructions, and catches Mr. McGraw on *a visit* to Philadelphia, serves a summons on him there, where, on a trial which took place, too much reliance was placed on a point of law, and a verdict was rendered for the plaintiff, without a trial on the merits.

In relation to the statement respecting one of our most able lawyers, and Mr. Winans' counsel, Mr. Meredith, folio 10, page 5, it appears from the affidavit of Mr. Tucker, president of the Philadelphia and Reading Railroad Company, on page 88 of defendant's No. 3, that they do not use the specified car, and that "Mr. Meredith was also the standing counsel for that company," from which it would appear that a friendly arrangement was entered into for the minimum sum of $500. Winans omits to set out these facts.

On page 7, folio 19, Winans further states, "that the delays herein detailed operated to prevent his obtaining remuneration for his improvement, and led him to apply for an extension of the term of his letters patent, under the act of Congress in such cases made and provided; which application was opposed, on behalf of the *New-England roads leading from Boston*, by the ablest counsel at Washington."

The record and files in the Patent Office would have shown Mr. Winans that such was not the case. The Boston and Providence Railroad Company was the only one represented, as the letter of the counsel will show; and the only objection made by the counsel was upon the point of law, that the chief clerk had no jurisdiction; and there seems to be a few affidavits now on file as to Winans not claiming the cars in use as being his invention, which were not referred to, at the time, by counsel. The statement in Winans' affidavit is put in such form, as to lead to the impression that he met with great opposition from all the New-England roads leading out of Boston, represented by the ablest counsel at Washington. The New-England roads from Boston numbered *six*, according to the affidavit of W. Raymond Lee, defendants' No. 3, page 113—folio 449, and only one of them opposed, in the manner stated. Mr. Winans had access to the record, as well as his counsel, and it was his duty not to swear by supposition, loosely making a general statement at variance with the facts. I will now leave Mr. Winans' affidavit, and refer to his partner and agent's, Mr. Gould's, to see what weight it possesses.

Mr. Gould, plaintiff's proofs, page 10—folio 35, states "that the said cars or carriages, which the said defendants have been and are constructing and selling, are constructed on the improved plan, upon the same principle, and similar to the improved cars or carriages mentioned and described in said complaint, and the schedule therein set forth, as patented to said Winans, and combine and embrace all the improvements invented by and patented to said Winans."

Mr. Gould is the only witness on the question of infringement. Mr. Winans himself is proved to have known how defendants' cars were built, yet he does not swear that the defendants' cars infringe his patent; but that he has been so informed by Mr. Gould. It is now proven, by numerous witnesses, that Mr. Gould's *judgment* is erroneous; and it is also shown that he is a party in interest in this suit.

I do not wish to call in question Mr. Gould's veracity, but merely to show the character of the phraseology of his affidavit. The question that we are trying is, whether the cars made by the defendants are on the same plan and principle as those mentioned in the plaintiff's specification; of which the court is to judge, and not Mr. Gould.

We say, therefore, that as Mr. Gould is an interested party, he is disqualified from being a competent witness; and that his is not a proper kind of affidavit, as it expresses a judgment, when it should contain detail of mechanical structure. Mr. Winans swears that the defendants infringe, only from hearsay, which comes from Mr. Gould, and these are the only two alleged witnesses on the subject of infringement, the one being the patentee, and the other a party in interest, whose name should have been affixed to the record in this suit as a party, coupled with that of Ross Winans.

I now propose to make some remarks upon plaintiff's witnesses and affidavits in relation to the Baltimore matter; and first, to show from them that Philip E. Thomas does *not know*, and in fact does *not profess to know*, and testify of his *own knowledge*, as to the machinery on the road, and the circumstances connected with it, but simply states his *information, understanding* and *belief*.

Mr. THOMAS,

In the Troy and Schenectady Case,	*In this Case,*
Page 15, cross-interrogatory 7 of defendants' copy, and page 7, cross-interrogatory 7 plaintiff's proof, says, in answer to the following question: "Was there any running gear composed of eight wheels used for transportation of *timber* or other articles on said road before the construction of the Columbus?" To which he answers: "*That none was used that he is aware of.*" In answer to the 6th cross-interrogatory, he says: "That he means to say, that in speaking of the eight-wheel cars, he refers to the running machinery. As to the body, a carpenter could do that. He refers to the running gear." In answer to cross-interrogatory 8th, he says: "That he has *no recollection* that he ever saw him (Winans) engaged in the construction of an eight-wheel passenger car; affirmant *had other duties*	In plaintiff's proof, John Elgar, Glenn, and others of the plaintiff's witnesses, swear, and the plaintiff's counsel here have expressly admitted before this court, that the *eight-wheel timber cars* were on the road in 1830, *the year before the Columbus was built.* In plaintiff's proof, page 16, folio 65, Mr. Thomas says: "That as affirmant *believes, in the year* 1831, *and in the early part of it,* the plaintiff *produced* a drawing of a car for passenger transportation, consisting of a body longer than the car body then in use, and mounted upon eight wheels. *Upon this plan* a car was afterwards built at the expense of the company in their shops; that it was called the Columbus," &c.

to perform, which prevented him from looking into Winans' manipulations."

In reply to cross-interrogatory 11, he states: "That he has *no recollection of said Conduce Gatch at all.*"

I now refer to the tenth cross-question and his answer.

Question.—"Have you a sufficient acquaintance with the construction of railroad cars, as of the running gear and bodies, *to designate differences of construction* in various kinds of eight-wheel passenger cars?"

To the tenth cross-interrogatory he answers, "*That he has* NOT."

1. Additional interrogatory:

"To what were your duties, as president of the said company, especially directed? Did they extend to the supervision of the department of machinery, or was that confided to others, and if so, to whom was it confided?" To the additional interrogatory by plaintiff, he answers: "That the details of the business of the company were divided into departments. The business of the company was complicated, and ramified itself into many different branches. *Affirmant's duty was to preside at the meetings of the board of directors; to receive the reports monthly, and submit them to the board; to attend to the general financial affairs of the company, and conduct the correspondence,* and to have a general superintendence of its concerns. To the best of his recollection, George Gillingham was the superintendent of machinery."

At folio 69. "And the passenger cars for the said Washington branch were directed to be built on the plan the plaintiff *recommended*, and which *is the plan of those at present in use on said road,* so far as this affirmant *understands and believes;* that the Washington branch was completed in 1835; and that the consideration and discussion, above referred to, must have taken place some time previous, though this affirmant is not now able to give the exact date of it; that when the Washington branch was opened in 1835, as aforesaid, the cars used upon it were those which, as affirmant *believes,* had been invented and perfected by said plaintiff."

I should violate my feelings of respect for that venerable gentleman, Mr. Thomas, were I to impute to him an intention, in the affidavit in this case, to mislead the court. He uses the language that he "*understands and believes,*" in relation to the drawing which now appears as the smoked drawing, and as to the construction of the Columbus and the Washington cars. He testifies in the Troy case that he has not sufficient acquaintance with the construction of cars, to designate differences of construction. He is not able therefore to distinguish between the smoked drawing and the Columbus, nor between the Washington cars and the specification; and his *belief* as to them, under such circumstances, is not evidence, and amounts to nothing against the positive testimony of Schryack and Shultz, who built them, that they are essentially different from the specification. He *never saw* Winans engaged in the construction of an eight-wheel passenger car, and possesses no *knowledge* as to what he was doing. He testifies that he does not recollect the timber cars on eight wheels, nor Conduce Gatch, showing most conclusively that he *knew nothing* of the *machinery*, nor *principal mechanics* who produced it. The plaintiff himself here proves that both the eight-wheel timber cars and Conduce Gatch were there in 1830, and thus shows that his own witness, Mr. Thomas, had not such knowledge of these matters of machinery and men as to make him a reliable witness in this case in relation to them.

It is also a noticeable fact, on the face of Mr. Thomas' affidavit, that he makes a clear distinction between his *belief* and his *remembrance*. At folio 65, p. 16 plaintiff's proofs, he states, in referring to matters not involving the issues here, and of no importance in this suit, that " the above facts are *remembered* by him;" and then he goes on to speak of matters of importance in this suit, connected with the machinery, but is first cautious to preface his remarks with the words, " that as affirmant BELIEVES, &c.," thus making a marked difference between *belief* and *memory*.

He also shows clearly, as he testifies, that he cannot designate differences in mechanics, or is not possessed of or has not cultivated a capacity to discriminate; for he testifies to a friction wheel and *box*, while Winans' patent of 1828 does not describe a *box* to the friction wheel, and the *box* inclosing the wheel was in fact invented in 1831.

Mr. Thomas' duties were the usual financial and corresponding duties of a president; he was not a mechanic, and hence cannot discriminate as to drawings or machinery. He possessed no *knowledge*, in fact, in the Troy case, and hence his assertions of *understanding* and *belief* are *not evidence*, and cannot have weight attached to them in this case.

George Brown is the next of the plaintiff's witnesses, to whose testimony I will call the attention of the court. He was the treasurer of the company in fact, and for a short time held the nominal position of acting superintendent of machinery. He does not swear that he is acquainted with machinery. He is a very respectable banker; and has been such since about 1809, in the city of Baltimore. There are discrepancies and errors in his testimony, and a general insufficiency of mechanical knowledge manifested.

He says, in the Troy Case, Page 7, in answer to interrogatories put by the plaintiff in person:

To the second interrogatory, he answers: "That he *thinks* the plaintiff was in the employment of the said Baltimore and Ohio Railroad Company, and to assist Mr. Knight in matters of machinery. Mr. Knight was chief engineer. *He cannot fix the date;* thinks it was in 1829 or 1830, on Winans' return from England."

3. "Do you know anything in regard to the *introduction* of the eight-wheel car into use on the said railroad, and whether the said plaintiff had anything to do therewith? State all that you recollect on the subject, and what opportunities you had of knowing anything about it."

To the third interrogatory, he answers: "That *with regard to the date he cannot speak;* but *one* thing he distinctly recollects — that he was the treasurer of the company *about* that time; and that Evan Thomas, Ross Winans and himself

In this Case, Plaintiff's proofs, page 20, folio 79:

"That he is acquainted with the plaintiff, and has been since on or about the year 1828; in the fall of which year the plaintiff visited the said city of Baltimore, having previously, in said year, obtained letters patent of the United States, to wit. on or about the 11th day of October, for a friction wheel and box."

At fol. 80. "That the plaintiff entered upon the duties of his said *appointment* (of assistant engineer of machinery) on or about the *first day of June*, 1830."

Folio 78. "That in the year 1831, in *addition* to his duties as treasurer, this deponent also acted as superintendent of machinery."

82. "That in the *winter* of the years 1830 and 1831, the plaintiff showed to this deponent a drawing of an eight-wheeled passenger car, or carriage."

83. "And a carriage was afterwards built and put on the road, in the *summer*

were talking over *machinery;* and whether it was suggested by deponent, by Mr. Winans or Mr. Thomas, he cannot say; but it was suggested by some one of them, *how much safer an eight-wheel car would be for* PASSENGERS *over the* four-wheel car; and Mr. Winans then *went on* to make a drawing of such an eight-wheel car as he supposed would answer the purpose."

4. "Look at the drawing, marked 'Oliver Cromwell, No. 1,' and say whether you ever saw it before, to the *best of your recollection,* and when, and whether it represents the car Columbus, *as you remember it.*"

To the fourth interrogatory, he answers: "That the drawing, marked 'Oliver Cromwell, No. 1,' represents the car Columbus, which *witness superintended the building of, or saw building every day* after Winans made the drawing. Deponent got the board of directors to have a car made after the drawing, as an experiment."

To the first cross-interrogatory, he answers: "That when Mr. Winans first came, he brought a plan of friction wheels; but *when he came into the company's employment, or in what particular capacity, he cannot say,*" &c.

To the fourth cross-interrogatory, he answers: "That after the drawing was completed, deponent never heard of any person mentioned as being the inventor of the eight-wheel car, except Mr. Winans, *which the suggestions of Mr. Evans Thomas, deponent, and deponent's father may have communicated to him in conversation.* What *deponent* communicated to him was this: that *the eight-wheel car* would be *safer* than the four-wheel car; and this led to *their* making a little drawing, of the manner in which *they supposed* it could be arranged, in a rude

of the year 1831, and its construction was *superintended by the plaintiff.*"

84. "That at different times alterations were afterwards made in said experimental car or carriage, and particularly in the running gear; *but he cannot now state with certainty what alterations were made,* but that each and all of the said *alterations* were made under the direction and at the suggestion of the plaintiff."

91. "And this deponent further says, that a copy of the drawing and plan of the car or carriage Columbus, the original eight-wheel passenger car or carriage hereinbefore mentioned, is hereto attached, and marked with the name of this deponent. *That the said copy has been carefully examined and compared with the original drawing and plan, and the same is correct.*"

In Troy Case,

Plaintiff's proof, pages 19, 20:

I now refer to the testimony of Jonathan Knight, chief engineer of the Baltimore and Ohio Railroad Company from 1830 until 1842, to prove that this statement of George Brown, that Winans was appointed assistant engineer of machinery about the first day of June, 1830, is erroneous.

I will read the third and fourth interrogatories, with their answers.

3. "Were you chief engineer of the railroad, constructed by the Baltimore and Ohio Railroad Company, during the years 1831 to 1836, both inclusive?"

Ans.—"I was chief engineer for the Baltimore and Ohio Railroad Company, as stated in my answer to the first interrogatory." (i. e. "I was chief engineer of the Baltimore and Ohio Railroad Company, from the commencement of the year 1830, till the 31st of March, 1842.")

way. The first suggestion to *deponent's* mind were two long beams of wood, in which the wheels were to work singly; a plan which was not adopted."

4. "Was the plaintiff, during any portion of that time, in the service of the Baltimore and Ohio Railroad Company? If yea, state when, to the best of your knowledge, his engagement commenced —and what were his duties in the service of said company."

Ans.—"The plaintiff returned from England to Baltimore, Md., in the year 1830; and by authority from the board of president and directors of said company, *I, as the chief engineer, appointed said Ross Winans to be my assistant in the department of machinery of said company. The appointment was made in the year* 1831, *and I believe took date from the first of that year.*"

The court will perceive that in Mr. Brown's testimony, he does not show any precise knowledge of either drawings or machinery; and wherever he is called upon to distinguish, he either says that he does not recollect, or, if he attempts, he makes mechanical errors. It is not to be expected that a respectable banker like Mr. Brown, whose mind is on his legitimate business, would be able to discriminate either in mechanical drawings or in machinery. He speaks of Winans' patent for a friction wheel and *box*. The patent is here produced, and it describes and claims the friction wheel alone. *There is no box in it.* The *box* inclosing the wheel was a subsequent invention.

He speaks of alterations in the running gear in the Columbus being made, but cannot describe them. He therefore cannot know how she was originally, or at first built; because he cannot recollect particulars of structure.

He swears, in the Troy case, that he cannot fix the date of Winans' appointment. In this, he attempts to fix it, but makes an error. He states, in the Troy case, that *he superintended the construction of the Columbus.* In this, he states that the *plaintiff superintended the same.* He states that the drawing was made in the winter of 1830–31, and that a *true copy* is annexed of the smoked drawing, which he *verily believes* is the original. The copy, on examination here, differs from the smoked drawing in the most essential particulars—that is, in not having any king-bolts for the trucks to swivel.

A great error is also made by Mr. Brown as to the time of Winans' appointment as "assistant engineer of machinery." He says it was about the first of June, 1830. Jonathan Knight, who made the appointment, swears that he appointed him his assistant engineer of machinery *in* 1831, *to take effect from the first of the year*. It was therefore a *nunc pro tunc* appointment; and although Mr. Brown, and others of the plaintiff's witnesses, namely, Elgar, Reynolds, Thomas and Woodville, state that Mr. Winans was appointed in June, 1830, still, the very man who made the appointment, and who was the chief engineer of the road, about whose business Winans was employed, asserts that it took place *in* 1831, and that it *dated* from the first of that year, *and therefore not in June*, 1830.

The reports of the president and directors of the Baltimore and Ohio railroad, for 1830, *made in October of that year, do not* mention Winans as being engaged in the service of the company. Mr. Brown and these other four witnesses have therefore antedated the appointment of Mr. Winans as assistant engineer of machinery; for as soon as Mr. Winans was appointed by Mr. Knight, namely, in 1831, the reports of that year make notice to that effect, and speak of the manner in which he was employed. On page 22 of the 5th annual report, in the 1st volume of reports of the president and directors to the stockholders of the company, dated Oct. 1st, 1831, it is stated: "In making these experiments, as well as in the arrangement of the machinery, I have been aided by *my assistants*, John Elgar and Ross Winans. *The latter gentleman is now engaged in planning the machinery and fixtures for the inclined planes.*"

This is signed by J. Knight, chief engineer, the gentleman who made the appointment, and who of course cannot be mistaken, as his deposition is corroborated by the report just read.

This statement also shows that Winans was NOT *experimenting with the Columbus, but that he was engaged in planning the machinery and fixtures for the inclined planes.*

Mr. George Brown was treasurer of the company, and a banker then and now; *financial* affairs, therefore, *pre-occupy his mind and memory*, and his testimony or *belief* appears *not* reliable as to *mechanical* drawings or *mechanical structure,* or the circumstances connected with their *details of arrangement or operation.*

I now ask the attention of the court to the infirmities, contradictions and discrepancies in the testimony of *John Elgar*, one of plaintiff's witnesses, apparently much relied on by him.

MR. ELGAR testifies,

In Troy Case,
Page 9 defendants' copy, page 12 plaintiff's proof, in these words, in answer to the third interrogatory:

"About the first of June following, Ross Winans returned from England, and came to Baltimore, and, *he believes*, entered into the service of the company, and was the chief of his time out at the depot of the company at Mount Clare. *Some short time after this, he, or some one, he cannot distinctly now recollect,* showed affirmant a drawing; *but he thinks it was a young man, by the name of Cromwell,* who was building or repairing the passenger cars. This drawing, *to the best of his recollection*, exhibited an eight-wheel passenger car, body and trucks together; and affirmant *understood, from general fame*, that Ross Winans was the originator of this drawing. Affirmant *does not* recollect that he had any particular conversation with Ross Winans on the subject then. Some time in the year following, *to the best of his recollection*, a car, which *he thinks* corresponded to the drawing, *was put on the railway*, and this car was called the

In this Case,
Page 24, folio 102:

"And this affirmant further says, that early in the summer of 1830, the said *plaintiff produced* a plan and drawing of an eight-wheel passenger car, that is, for a long bodied car resting on two bearing carriages of four wheels each; that he has seen to-day what *he believes* to be the *same drawing,* a copy of which is hereto annexed, marked 'John Elgar, A,' which plan was approved by the directors of said company, and a car ordered to be built therefrom, at the said shops, under the direction of the plaintiff."

CROMWELL himself, at page 45, folios 206 to 208 of plaintiff's proofs, says:

"That in the month of *June*, 1830, the *plaintiff* was engaged by the railroad company, as this deponent was *at that time informed, as assistant engineer,* to aid in getting up the required machinery of the road, and, together with John Elgar, also an engineer of machinery, and the chief engineer, Jonathan

Columbus; *he thinks so.* Affirmant rode in it *once* to Ellicott's Mills that summer —the summer of 1831. In 1832, affirmant left Baltimore, and was engaged on the Columbia railroad, Pennsylvania."

In answer to the fifth interrogatory, Mr. Elgar says: "That he has looked at the drawing, marked 'Oliver Cromwell, No. 1,' and attached to the deposition of Oliver Cromwell, and has read the specification of the plaintiff's patent. He does *not consider the car Columbus,* SUPPOSING it to be truly represented on said drawing, *fulfils the requisition of said specification.* He thinks the specification calls for the axles and wheels of each truck to be placed *as near to each other as may be,* to work freely; and that each truck should be separated from the other, under the body of the car, *as far apart as the strength of materials and convenience will admit;* and on this drawing, the *wheels of each truck are farther apart than is called for in the specification, and the two trucks* are placed nearer together, or *farther from the ends of the body, than is called for in the specification.* He thinks the differences he has stated are important. *He cannot* say that the drawing now shown him is the one shown him by Oliver Cromwell; but said drawing is a fair representation of the eight-wheel car Columbus, *as he now recollects it.*"

On page 13, defendants' Troy case, page 15 of plaintiff's proofs:

1. "In answer to the first interrogatory, you say you understood that Mr. Winans was employed as an assistant engineer of machinery, in 1830, by the Baltimore and Ohio Railroad Company; can you say, positively, that he was so employed, and why you have that impression?"

To the first cross-interrogatory, he answers: "That he cannot state it posi-

Knight, was frequently at said shops, which were situate at what is known as the 'Mount Clare Depot;' that *early in the year* 1831, George Brown, Esq., then treasurer of the company, was in the carriage-makers' shop, and while there, commenced a conversation with this deponent about the road and the carriages in use thereon; that in that conversation he mentioned the name of the *plaintiff, who he said was getting up a drawing of a passenger carriage upon a new and improved plan,* and which was different from any carriage at that time in use; that *shortly afterwards, and within a few days, a drawing and plan of an eight-wheeled passenger carriage* was brought to the carriage-makers' shop by the *plaintiff,* and handed to *this deponent* by *him,* with *instructions* that *a carriage should be built therefrom ;* that he has this day seen the same drawing, and he is positive as to its identity, *a copy of* which is hereto annexed, signed by this deponent. *The drawing shows the body and the running gear, as the carriage was built. The drawing was handed to this deponent in the month of February or March,* 1831, and the *work* upon the body was *begun* within a day or two *afterwards.*"

210. "That there was but one drawing and plan," &c.

Mr. Thomas, page 16, folio 65, plaintiff's proofs, says: "That, as affirmant believes, in the year 1831, and in the early part of it, the plaintiff produced a drawing of a car for passenger transportation."

George Brown says, folio 82 plaintiff's proof: "That in the *winter* of the years 1830–1831, the *plaintiff showed to this deponent a drawing of an eight-wheeled passenger car or carriage,*" &c.

Michael Glenn says, at page 36, folios 162, 163, plaintiff's proof: "That the

tively, but he understood *from Winans that* soon after his return he *was* taken into the company's employ—within a month, *or two* or *three* months; he *thinks* he *understood* this from *Winans;* he cannot be positive about Winans being an assistant engineer at that time; but, affirmant is positive that he was an assistant engineer in 1831; *he heard Winans say so in* 1831, *or the spring of* 1832, and he heard the chief engineer, Jonathan Knight, say that Mr. Winans was his assistant engineer of machinery. Affirmant was at that time assistant engineer of machinery; and when Winans first came there he thinks Winans was assisting affirmant, and both of them assisting the chief engineer; *and shortly before he went to Pennsylvania* he heard Jonathan Knight say that Winans was his principal assistant engineer of machinery."

8. "How often did you see the drawing, which you say was shown you by Cromwell, of the car Columbus, and do you know that the running gear of the Columbus was built after that drawing?"

To the eighth cross-interrogatory, he answers: "*That he does* NOT *recollect to have looked at said drawing more than* ONCE. He may have done so. He does *not know that the running gear of the Columbus was built after that drawing.*"

The report of James P. Stablee, superintendent of construction, dated September 30, 1832, page 102 of 6th annual report, volume 2, states:

"I feel bound also to acknowledge the important aid derived from *John Elgar,* engineer of machinery, who *has been engaged in superintending the construction* of the *turn-outs, from the commencement of the work until July last.*

"The superiority of his improved turn-out (a drawing of which was given first car that he ever saw, to which the name of 'a double-truck, long-bodied car' could be applied, was the eight-wheel car, Columbus, the invention of the plaintiff, built at the shops hereinbefore mentioned, *commenced in March,* 1831, *and finished on or about the* 1st *day of July,* 1831. That the facts and circumstances connected with the building of said car, Columbus, are as follows: that, *during the month of February, preceding the commencement of said car, Columbus,* or about that time, this deponent heard it spoken of, in the shop where he worked, *that the plaintiff was getting up a new kind of car,* and that a plan or drawing, on paper, showing the invention, would be brought to the shop in a few days." "That, *within a few days after* this deponent heard that the plaintiff was getting up a new kind of car, *the plaintiff* brought to the shop a plan or drawing, showing a side elevation of an eight-wheel passenger car," &c.

John Elgar, also, who testified in the Troy case that he does *not* recollect to have looked at said drawing *more than once,* and that it then was in *Cromwell's* possession, in this case says, folio 104 plaintiff's proof: "That the said drawing was *often* seen by this affirmant, in open vie w at the shop of Mr. Cromwell, while the said car was building, viz: in the spring of 1831. *The drawing and car are as familiar to this affirmant as any event connected with said road.*"

in the last annual report) over any other as yet in use, and his *indefatigable exertions in bringing to its present perfection this important improvement in the machinery* of the Baltimore and Ohio railroad, deserve the highest commendation.

"Much credit is also due to Amos Farquhar and *Ross Winans*, for the very efficient manner in which their services were rendered during the months of October and November, 1831, in the *procurement of materials and the erection of the chambers* for the machinery at the inclined planes.

"Signed,
"Respectfully,
"JAMES P. STABLER."

This Mr. Elgar, in his testimony in this case, states that he had the superintendence of the department, including the work-shops, and he " gave *his directions*, for the construction *of machinery to be put on the road*, to said Gatch." What were *his directions*, and to *what machinery* did they refer? He is proved by the reports to have been engaged personally, from the commencement of the work to July, 1832, with his *turn-out;* and the fair, legal inference therefore is, that *his directions* referred to patterns for this, his invention, the object of his " indefatigable exertions," and primarily, if not almost exclusively, occupying his mind. While Winans was planning machinery for the inclined planes, and experimenting to improve his friction wheel, Elgar attempts to make it appear that Winans was experimenting on the Columbus; he says, " that *he has been familiar* with the said *experiments* since the building of the Columbus, and the several cars known as the Winans eight-wheel cars, upon plans which, though differing from said car Columbus, were built with a view to improve the idea attempted to be developed in that car, as hereinbefore mentioned. That the said cars, and each of them, *as he fully believes*, were the invention of said plaintiff, and were so spoken of *at the time of their construction;*" " that he left the service of the said company *about the first of July*, 1832, and entered that of the *Columbia Railroad Company, of Pennsylvania*, in which service he *remained about two years."*

It is proved by the reports, and the witnesses on both sides, that the Winchester, Dromedary and Comet were *constructed* in 1833–34, and were all in use in the summer of 1834. Mr. Elgar had been *absent in Pennsylvania;* yet he professes to *believe* that he has the capacity to discern, from the State of Pennsylvania, minute particulars as to plans, persons, remarks and mechanical changes, if any, going on in the workshops at Baltimore. Such pretensions or assumptions are incredible. The report of Oct. 1, 1831, by Mr. Knight, speaks of the manner in which both Mr. Elgar and Mr. Winans had been employed, the experiments that they were engaged in, and yet nowhere, in any of said reports, is there the slightest allusion to Conduce Gatch's car, " Columbus," and not the slightest pretence that Ross Winans had either invented or was engaged in experimenting on that or any other eight-wheel car. Had he been thus occupied, a report which narrated the doings of Mr. Elgar and Mr. Winans would certainly have made, at least, allusion to it; but such a structure as an eight-wheel car is not even hinted at, and yet both sides of this case admit that the " Columbus" at that time existed, and the newspapers of June and July, 1831, speak of her as " a capacious new car, on eight-wheels."

Mr. Elgar, signing himself " assistant engineer," makes a report, dated 30th Sept., 1831, at pages 100, 101, 102, of the 5th annual report of the 1st volume of reports, and addressed to Jonathan Knight, chief engineer of the Baltimore and Ohio road, in which he names experiments, made in connection with the motive or steam power, and the machinery of the road; but he does not make the slighest mention of experiments with eight-wheel cars by Winans.

If what Mr. Elgar states in his affidavit be true as *facts, not* as *his belief only*, why did he omit it in his report of 1831? And more especially, why, as he and Mr. Knight were in the habit of reporting all that Mr. Winans did?

Mr. Winans' name is expressly connected with certain experiments, specified in these reports, to show how the chief engineer and his assistants were engaged; but strange to say, if these be true statements *in fact*, not a syllable is lisped, not a word uttered, which would imply that they had even noticed that an eight-wheel car was on the road, or which would compel or even suggest the inference that he, Winans, was making improvements on the eight-wheel car.

Also, the fact is shown that Mr. Elgar himself was engaged in making experiments with chilled boxes of his invention, and other matters which are here enumerated, displaying an intention to show how their time and minds were occupied; yet neither he nor Winans appear connected with, nor take any notice of Gatch's eight-wheel car.

In the report of James P. Stabler, it is further shown that Ross Winans, in the months of October and November, was engaged in procuring materials and constructing the chambers at the inclined planes. He had been previously during the year engaged in planning and constructing the *machinery itself*, and now we find him constructing chambers at the planes to apply the machinery. The occupation of Mr. Elgar and Winans, during 1831 and 1832, is fully shown by the reports made at the time, and by our testimony; and hence Mr. Elgar is directly contradicted in his assersions as to Winans inventing and experimenting with the eight-wheel car.

The plaintiff's counsel have said that Mr. Elgar's statement in his affidavit, p. 24 plaintiff's proofs, just now referred to, carries the invention back to the summer of 1830. But we have already shown, by Mr. Knight's testimony, that Mr. Elgar is mistaken as to the date of Winans' appointment as assistant engineer; and if he be mistaken as to one date, the strong probability is that he is also mistaken in relation to the other.

In his affidavit in this case, also, he states that the *plaintiff produced* the plan; and that it was often seen by him, in open view, at Cromwell's shop; and that the *drawing* and *car are as familiar to him as any event connected with said road*. While in the Troy case he swore that he thought it *was Cromwell who showed him the drawing;* that he had *no conversation with Winans on the subject;* that *he did not recollect to have looked at the drawing more than* ONCE; and that he did *not know* that the running gear of the Columbus was built after the drawing.

And Cromwell himself, in this case, page 45—folio 208, says: "*That the drawing was handed to this deponent in the month of February or March,* 1831, and the work upon the body was begun within a day or two afterwards." Mr. Thomas, Mr. George Brown, and Michael Glenn, who are plaintiff's witnesses, and also all the defendants' witnesses, fix the first production of a drawing, in point of time, *in the latter part of February or early part of March,* 1831.

That Mr. Elgar is wrong in the date, both in the year and time of the year, as to the first production of a drawing, is clear, both from his testimony as to seeing it only once, and then in Cromwell's possession, and the testimony of Cromwell, Thomas, Brown, Glenn and all defendants' witnesses on the subject, who fix the time about the first of March or latter part of February, 1831; although the *belief* of the plaintiff's witnesses differs from the *knowledge* of the defendants' witnesses as to what was, and what was not, on the drawing. They agree as to *time,* and that Winans *produced* it. We are therefore led to the undoubted conclusion that Mr. Elgar is in error as to the *time.*

His assertion as to "general fame, that the plaintiff was the originator of the drawing," I will remark, is not sufficient ground to admit of his testifying to that as a fact, and his *belief* is not evidence. He only thinks that he saw a drawing in Cromwell's possession, and saw it *once only,* and he understood that Winans produced it; and he only rode *once* on the Columbus. He is not, therefore, *familiar* with either the drawing or the car. What his familiarity with other events may be, with which he designs to compare familiarity with the drawing and car, does not appear; old age and the vicissitudes of life have probably clouded and confused his mind and memory. His testimony is neither reconcilable nor reliable.

I will recall the fact that he, in this and the Troy case, totally contradicts himself as to the number of times he saw the drawing, where he first saw it, as to the Columbus corresponding with it, and as to his opportunity of knowing the peculiarities of the eight-wheel cars in use. His affidavit in this case is drawn in the most exaggerated terms, as compared with his own words in the Troy case; and he swears to facts *on belief* now, that he did *not know* before. This phraseology, then, of Mr. Elgar's affidavit cannot be relied upon, as it is contradicted by himself and others so flatly and materially.

In the Troy case, no pretence was set up by the plaintiff to a drawing before the beginning of March, 1831, and here it is evidently with the intent of trying to go behind the eight-wheel Allen engine, and the eight-wheel Wood cars, and the eight-wheel Trussel cars, in point of time; but the truth that these cars existed before the drawing was made, is too apparent, from the testimony, to require any further remarks on Mr. Elgar's affidavit.

The next of the plaintiff's witnesses, to whose evidence I shall bring the attention of the court, is *Mr. Woodville*, a respectable gentleman no doubt, now an exchange broker. He was auditor and superintendent of transportation of the Baltimore and Ohio railroad.

He says, in substance, that Mr. Gatch had no right to invent; it was not his business to invent, or reflect, or to think, or possess a knowledge of the business of the road, or improve in the machinery of the department of which Gatch had charge; although he, Woodville, as if disposed to credit himself, lays claim to the merit of an invention in the backs of seats, without stating or describing particularly what he did, and in what mechanical peculiarity his invention consisted. *He* certainly was not appointed to invent; he was superintendent of transportation and auditor *only*. Whether the right to invent was a prerogative of his offices, or not, he does not explain.

Notwithstanding the pretensions set up by this gentleman over the "workmen," as he has been pleased to term the other respectable gentlemen who have testified for the defendants here, after denouncing them and their rights and privileges, we will now proceed to show that this gentleman, Mr. Woodville, not only was not, from the nature of his occupation, qualified to know much of the details of mechanical structure and principles, but he does not even know when he was appointed superintendent of transportation.

He says, on page 31, at folio 136 of plaintiff's proofs: "That he was formerly, and until the year 1836, an officer of the Baltimore and Ohio Railroad Company, as master of transportation, and continued as such *from the opening of the road, in the month of May,* 1830, until the month of July, 1836."

In his own report, dated October 1st, 1831, at page 128 of the first volume of reports, he states:

"Having, in addition to my duty as auditor, been appointed superintendent of transportation on the railroad, I respectfully submit to you the accompanying documents, *as the result of the operations in that department, from the 1st of January to the 30th of September, of this year.*"

His report, made at the usual time of making the reports, the first of October, instead of embracing the transportation from the 1st of October, 1830, or from May 22d, 1830, when the road was opened, embraces only from the 1st of January, 1831, showing that he was appointed superintendent of transportation the "1st of January, 1831," and *not* "in the month of May, 1830." He previously was appointed auditor. Evidently, his were office or book-keeping duties, not mechanical either in invention, construction, nor in observation. This proves that he is in error as to dates; and a fair inference is, that he is also in error as to the time that he first observed the eight-wheel wood cars, if his capacity for observation in mechanics, at an early day, was sufficient to distinguish them from the eight-wheel timber cars; and if he had reflected a moment, he would have remembered or ascertained that another gentleman, since dead, took charge of the transportation in 1830.

In order to show the *aristocratic tone* of his testimony, and its unreasonable character, particularly in the courts of this republican country, where, I trust, such feelings will never find support, I refer to page 32—folio 142 of plaintiff's proofs.

Mr. Woodville says: "*That the said Gatch had no connection whatever with said road, and had no authority to get up, or build, or put a car or cars on said road, or any control of any kind thereon;* that he was a carpenter in the company's shops; confined to his shops, without control in any way, beyond the limits of such shops; and that the said Rupp was a carpenter, Eichelberger a painter, and Forrest a blacksmith, neither having a control of *nor a knowledge of the business of said road;* that each and all were *workmen.*"

As to this, I have but to remark that Mr. Woodville probably knew nothing but the state of his transportation and audited accounts; he pretends to know, and swears to entirely too much, when he attempts, as he has, to swear four of the most respectable, intelligent and skillful mechanics and gentlemen of Baltimore into utter ignorance of things daily seen by them; and the pay rolls produced by plaintiff show exactly what Mr. Gatch's position was. He was master of the car building; Woodville was master of transportation, and auditor. Gatch controlled and directed his department, and was entrusted with large sums of money to defray its expenses; Woodville reports as to the "*commerce*" of the railroad. Gatch *plans and builds cars* to carry the *freight;* one is no more an officer, and no higher in position, in fact, than the other. Mr. Woodville, with all this manifest feeling, error as to time, and want of knowledge in mechanical structure, is certainly not reliable as a witness on such points. His mind must have been occupied with his audits and transportation accounts. All of the plaintiff's witnesses have spoken of "burden cars," as such; which are distinct from the eight-wheel wood and trussel cars, and for a different purpose. As an instance of this, I will call attention to the affidavit of Mr. B. H. Latrobe. At page 34—folios 149, 150 of plaintiff's proofs, he states: "That in the spring of the year 1831, this deponent was an assistant engineer, in the service of the Baltimore and Ohio Railroad Company, *and was engaged in making drawings of the cars and machinery in use on said road.* That he is very confident that had there been an eight-wheel car for *burden* or *passengers* then in use, he would have known the same and made a drawing thereof, as his object was to represent the road and its machinery, with all its improvements up to that time, in detail. The drawing, marked 'B. H. L., No. 1,' represents the *only burden cars* then in use, which were four-wheeled cars; the form of body and minor parts differing in some cases."

By the report of Jonathan Knight, dated Oct. 1st, 1831, at the end of page 22 of 1st volume of reports, in the 5th annual report, we prove that this drawing was a *flour car*, to haul *flour barrels alone*, and not intended to represent the passenger, platform, eight-wheel wood and trussel cars, or any other car, of all which no drawings were intended to be given. It was drawn and presented, in the report, simply as a convenient car to haul *flour in barrels* to Baltimore. Taking this as an example of the legitimate meaning of the use of the words "burden cars," which is frequent in the plaintiff's testimony, it does not refer in any way to the wood and trussel cars in controversy here, and all that testimony is therefore equivocal and irrelevant.

I will now refer to the testimony of *Michael M. Glenn*, on whom the plaintiffs seem to rely, to show that he contradicts himself, and in fact is entitled to no weight; that he really recollects nothing that is true and material in this matter.

He testifies,

In the Troy Case,	*In this Case,*
Page 33 plaintiff's proof, and page 28 defendants' Troy case:	Plaintiff's proofs, folio 165, &c., Glenn says:
"In 1830, or 1831, I first saw a drawing of an eight-wheel car. It was brought to the shop, I believe, by Winans, and laid on Cromwell's table; it is the one shown to me, marked 'Cromwell, No. 1.' *I first saw it on Cromwell's bench; I don't remember hearing it previously spoken of.*"	"That within a few days *after this, deponent heard* that the plaintiff was getting up a new kind of car. The *plaintiff* brought to the shop a plan or drawing, showing a side elevation of an eight-wheel passenger car." "The bearing carriages were shown on a plan or drawing, *corresponding in all respects to the four-wheel carriage at the time in use upon the road for passenger cars;* and the only carriage which had
"I assisted building the trucks of the Columbus, in the shop adjoining Cromwell's. *I think* the trucks were built	

while the body was being built from the drawing. *Jacob Rupp and I were the workmen. We considered they were built under Winans' direction." " The trucks were built in the form shown in the drawing."

" *The Columbus was twenty-seven feet long.*"

" I never knew an eight-wheel car built before the Columbus, after *that plan. I had never met with any such truck before that, or knew of any.*"

On cross-examination, he says: " *The first time I saw the drawing, it lay on Cromwell's bench. I did not see it brought there.*" "*I did not have the drawing before me when we worked at the trucks of the Columbus. Conduce Gatch gave us directions,* but not more than the plaintiff. Mr. Gatch directed *us to go on and make the trucks.* We worked from the drawings on a board. I think Mr. Gatch made that drawing." " I think Winans saw the drawing on Gatch's board. I have no doubt he saw it."

been known or used upon the road *for that purpose,* up to and including the invention of the Columbus." "The plan or drawing was projected for three-quarters of an inch to a foot. The length of the drawing upon the paper *was and is* eighteen inches; and *was and is* eight inches in height, to the top of railing, which *would give and did give* a length of *twenty-four* feet to the body."

" *That during the building of the Columbus, the plan or drawing was in open view in each of the shops,* although principally in the carriage shop."

" That the plaintiff was at the shops nearly every day; and he *believes* that he may truly say, frequently many times a day. *That he gave* ALL *the directions about the building of the car Columbus.*"

" And this deponent further says, that there were not any trucks *built for use, or upon said road, in any way, with a bolster piece or bolster pieces extending across the frame of the truck, or with each truck having four wheels, until the year* 1834."

Jacob Rupp himself contradicts this Mr. Glenn, and testifies, at defendants' No. 2, page 15, folio 59, &c.:

" The next car built on this principle was the Columbus, intended to carry passengers for fare on the road. This car was built in 1831. She was commenced in March, 1831, and finished and put on the road by the 1st of July, 1831, and on the 4th of July, 1831, she ran to Ellicott's Mills. *I myself, with the assistance of Francis Gatch, now deceased, built the trucks and bolsters of the car Columbus, under the instructions and directions of Mr. Conduce Gatch, and I also put the cornice around the body of the car.*

" *I did this work altogether by Mr. Gatch's instructions.* To direct us to make the trucks, he, Mr. Gatch, made a drawing on a board, giving the dimen-

Jacob Rupp, to whom Glenn so confidently refers, also contradicts him in defendants' No. 4, pages 60, 61.

He says: " I have examined the model now before me, marked ' defendants' model, K,' and signed by me. It is a true model or representation of the car Columbus, as built at the Mt. Clare depot in the spring, and finished about the 1st of July, 1831. The trucks then *had plain boxes* on, and were constructed like those in the model I built them, *with the assistance of Francis Gatch, by Conduce Gatch's directions.* This car also had on her, when she first went on the road, *iron baggage rods, like the model;* they were made and put on by Leonard Forrest. The Columbus had nine or ten sliding windows on a side, each about two feet square, and twenty-seven or thirty panes

sions of the timbers of the trucks and the bolsters; and we bored the holes for centre pivots or bolts, on which the bolsters turned, by Mr. Conduce Gatch's instructions and directions. Reuben Naylor, now deceased, cut the stuff for the trucks out of the rough, by Conduce Gatch's directions. *No other persons worked upon making the trucks and bolsters of the car Columbus.* The centre bolts to go through the bolsters, and all the other iron work, except the axles and wheels, were made by Leonard Forrest, under Conduce Gatch's instructions. I saw Mr. Gatch give him these instructions to make the centre pin, and other iron work. When the Columbus was first put on the road she had a plain top, without any railing; after she had been on the road between one and two months, Leonard Forrest made an iron railing and put on the top, and seats were put on the top for passengers to sit on. This iron railing was designed after the Columbus was put on the road. I saw the drawing given by Conduce Gatch to Oliver Cromwell, and saw it afterwards in Oliver Cromwell's possession. It was intended to represent a car body without any upper bolsters, and without any railing on top, and there was not any truck or other running gear on that drawing. The drawing was on paper, and marked with yellow paint to show the frame. Oliver Cromwell was instructed in building the body of the Columbus by Conduce Gatch. I worked in the next shop; Cromwell also had the drawing in his shop; it was a common sketch, and there were no dimens ons of the frame on it,—not that I saw. There were no alterations made in the distance apart of the wheels of the car Columbus. I had charge of that part of the work, that is, the trucks. The bo sters were of glass, of small size, about seven by nine, over the sliding windows on each side. John Rupp, my brother, made the sash. I put the cornice around the body of the Columbus, under the eave. The model K is a true model of the Columbus, as she was first built, in every particular, upon Conduce Gatch's plan and directions.

"I have examined a lithograph drawing, marked or printed on the face 'Columbus,' and in the lower right-hand corner 'Lith. of Rich'd A. Pease, Albany.' *It is not at all like the drawings from which the car Columbus was built, and is not like the car Columbus when she was built.*

"This drawing or lithograph I have placed my signature on, to identify the same. It has a *wooden railing on top,* with the uprights continued up above the top side pieces, to form the uprights of the railing, and has only six sliding windows on a side, with eighteen panes of glass. The Columbus *never had a wooden railing on top,* and the sliding windows and panes of glass were not so few in number nor so large in size; they were more in number and smaller in size. The Columbus also had a door at one end, and a pair of steps, and a panel at the other, on each side, like the model. This lithograph has two doors on a side, one at each corner. *The trucks on this drawing are not such as I built for the Columbus; the finish and manner of putting the truck frame together is not like the trucks made by me for the Columbus in* 1831. This drawing has Winans' friction wheel boxes on the trucks; the friction wheel and boxes were afterwards put on, in 1832, and required double side pieces, being a different construction of truck from those put under the Columbus when she was first built, in 1831. The trucks then had single side pieces, with

made of locust wood; and sometime after she had run on the road, I put stouter bolsters under her, to carry a heavier load; the first bolster would spring with a heavy load, but the wheels were not altered, and were ten or twelve inches apart, between the flanches, about the same distance they are used now on the Baltimore and Ohio road. *Mr. Ross Winans never gave me any instructions in building the Columbus, and I do not know of his giving any to Oliver Cromwell. I did not know, or hear, that I recollect, of Ross Winans at that day.* The next eight-wheel car built was the Winchester, and the next was the Dromedary; all these cars were the same, in principle of construction and operation, as Conduce Gatch's cordwood and freight car, with permanent frame, put in operation on the road in November, 1830."

plain boxes, and the perches had hownes and were put together like the model K. *I recollect distinctly, for I did the work, assisted by Francis Gatch.* The ground plan of body framing shown in the lithograph is not at all like the Columbus. She had four long pieces to sustain the floor, and the two side pieces extended beyond the end pieces, all as shown by the model K. This lithograph has the side pieces flush with the end pieces, and no long middle pieces to sustain the floor; it has one cross piece in the middle, extending across in the same direction as the floor boards shown at the end, and therefore could not sustain them.

"Among the many eight-wheel cars built by me for the Baltimore and Ohio Railroad Company, and which I assisted to build, were the trussel cars, on eight wheels, or two trucks of four wheels each; of which great numbers were built for carrying carriages and horses, and wagons and stock, and such freight. The first one or two of these trussel cars for carrying carriages and horses altogether harnessed up, in them, was to the best of my recollection built and used late in the year 1830. I think in December, 1830, it was used to carry the carriage and horses of some distinguished person out to the Relay House, sometimes called the Half-way House. I do not recollect the name of the person. This trussel car had two four-wheel trucks, with bolsters to swivel under the body, the same in principle as the eight-wheel wood cars then in use, and the same as the trussel and other eight-wheel cars now in general use. The ends of the trussel or frame let down, with hinges, for the carriage and horses to go up into the car. It was about twenty feet long; long enough for a carriage with the horses attached to stand up in, and be transported along the road. Numbers of such trussel cars were built in 1831, and used for transporting stock, hogs and cattle, when the road opened to Frederick, in the latter part of 1831. They were placed upon two four-wheel trucks which swivelled the same in principle as the eight-wheel cars now in use generally. These one or two eight-wheel trussel cars, and the eight-wheel wood cars, of which I have before testified, were all on the same principle, and were in use on the road in 1830, before the car Columbus was commenced; she was commenced in March and finished in July, 1831, like the model K."

Michael Glenn swears, on page 41 plaintiff's proofs, folio 184: "That Francis A. Gatch was a workman with Oliver Cromwell; *that he did not work in the carpenters' shop at any time, nor did he assist in making the bearing carriages, or trucks, or bolsters of the car Columbus, nor any part of said trucks or bolsters.*"

Glenn also swears, page 35, folio 157 plaintiff's proof: "That he was employed by said *John Elgar in building turn-outs and other work upon the line of said road*, but that he frequently returned and was employed in said shops; that sometime in the year 1834, the said Gatch left the employment of said company, and did not return again."

Mr. Nathan Randall, page 72, folio 280, in speaking of this Michael Glenn, testifies: "I conversed with him in the presence of Mr. W. W. Hubbell about his testimony in the Troy and Schenectady case. He denied swearing, in that case, that the Winchester drew by the body. He said that it was impossible for him to recollect about the cars; his mind had been on other pursuits,—he had been farming for about twelve years, and had not had his mind on the subject of cars. On seeing the model K of the Columbus, he voluntarily said, 'why, that is a model of the old Columbus! here is the perch; I recollect it and the iron railing on top.' *He said that when in the employ of the company, his chief employment was in laying down switches; that he knew more about laying down switches than anything else; that he could not recollect much about the cars,—it could not be expected of him.*"

It also appears on page 7, defendants' No. 4, folio 272, that he was agent for Winans in Baltimore. Edward May states: "Michael Glenn first came to me, and asked me to tell him all about what I knew of the eight-wheel cars. I was not acquainted with him; and in a few days afterwards Glenn came again, and brought Chas. D. Gould with him."

This Michael M. Glenn shows conclusively that he is not to be believed in his testimony. He contradicts himself as to knowing and not knowing that the plaintiff brought a drawing to the shop. He says, in this case, he did not see it brought there. In the Troy case, he says it was brought to the shop, I believe by Winans, and laid on Cromwell's table.

The fact is, Gatch brought the drawing to the shop; he received it from George Brown, to whom Winans produced it.

In this case, Glenn makes the Columbus 24 feet long; in the Troy case, he swore she was 27 feet long. She has therefore shrunk three feet in his mind, as he now testifies on a smoked drawing before him, assuming it to be the original, and assuming that the Columbus was built from and exactly like it.

He testified, in the Troy case, that Jacob Rupp and he built the trucks, and considered it Winans' invention; Jacob Rupp himself has since been found, and he swears that Francis A. Gatch and he, Rupp, built the trucks, and Winans gave no directions, and had nothing to do with them; and that the smoked drawing, of which the lithograph is a copy, is not like the drawings from which the car was built, and not like the car when she was built; the differences he enumerates show it impossible to have built the car from such a drawing.

Francis A. Gatch worked on both the body and the trucks; Glenn testifies that he did not work on the trucks.

These facts are corroborated by Conduce Gatch, Forrest and Eichelberger.

Glenn's statements are flatly contradicted; he professes to know entirely too much.

His admissions to Mr. Randall, and his own statement, show that he was principally engaged in laying down turnouts with John Elgar, who it will be recollected was so employed from the commencement of the road until July, 1832. Hence Glenn, not knowing the facts as to the work at the shop, contradicts himself and the other witnesses, and makes a statement which is inconsistent, as it would be impossible for him to know what was going on in the shops, when he was away on the unfinished portions of the road, constructing turn-outs.

The court cannot fail to notice that he contradicts himself, also, directly as to the trucks. In the Troy case, he seems to have recollected that the Columbus had thorough brace bolsters, and swore he never saw such trucks before. In this case, he is apprised that that will not do, as now he finds, on looking at the smoked drawing, that it does not agree, and hence he swears directly the reverse; he swears that the trucks were the same as in common use.

In the Troy case, he swore that Gatch gave directions to build the trucks. In this, he swears that Winans gave all the directions.

The truth is, Mr. Glenn *knows* nothing in fact of the particulars of this matter, and his statements cannot be believed. His *belief* is not evidence.

I now turn to *Oliver Cromwell's* testimony, one of plaintiff's witnesses, who smoked and produced the drawing, at the trial of the Newcastle and Frenchtown case, to prove that his story is unfounded, and improbable on its face; and that he is contradicted, and not entitled to weight.

At page 44 plaintiff's proofs, folio 202, he states, that: "This deponent was employed by the *superintendent of machinery* of the Baltimore and Ohio Railroad Company."

I now turn to Gatch's evidence, *plaintiff's proofs*, page 103, folio 491, where, in answer to the question as to what was his position and duty, &c., he says: "*I was called the master carpenter; superintended building cars, of various kinds; purchased all the materials, and had the supervision of all the men employed in the building of such cars. All the cars that were built at the company's expense were built under my direction* (except a few passenger cars), during the years 1830 and 1831."

The court will observe that Cromwell states, in his affidavit, that he was employed by the *superintendent of machinery*.

I now turn to voucher No. 32, page 106, plaintiff's proofs, where it will be further seen, from the following pay-roll, that Conduce Gatch was *master carpenter*, and employed, directed and paid Cromwell: "We, the undersigned, do acknowledge to have received from *C. Gatch, master carpenter*, the amount hereunto set opposite our names, respectively, in full payment of our services for the time herein specified, while *employed under his direction* on duties relating to the construction of wagons, &c., for the Baltimore and Ohio Railroad Company, November 30th, 1830."

At No. 8, of the list of names of the undersigned, we find:

"No.   Names.   Occupation.
" 8.   *O. Cromwell,*....*Coachmaker, &c.,*...*Signed, O. Cromwell.*"

Showing, by his own signature and acknowledgment, that he was under the employment and direction of Conduce Gatch, and was paid by him.

At page 104, plaintiff's proofs, folio 498, in reply to a question, Mr. Gatch says: "Yes, sir; *I paid them for all the work done about the making and repairing of cars, whether done in or out of the shops, and the time was included in the pay-rolls I returned.*"

It appears to us from the statement in Mr. Cromwell's affidavit, "that he was employed by the superintendent of machinery," that it is intended to leave the impression on the mind of the court that he was a principal man or conductor, and that Mr. Winans, therefore, handed the drawing to him, as he testifies. But the fact, as shown in Gatch's testimony and the plaintiff's pay-rolls, that he was under his employ as a *journeyman*, proves that he (Cromwell) was not the person, in all probability, to whom it was given. He was a journeyman; and therefore not the person to whom a drawing, with orders to build a car, could or would properly be given on its first presentation.

In plaintiff's Troy case, also, page 29, Cromwell himself says that he "commenced working for the Baltimore and Ohio Railroad Company in 1830, as a *journeyman;* Conduce Gatch was foreman of the shop." He was not, therefore, a principal, as has been stated in one of the plaintiff's affidavits, and not the person to whom Winans could give a drawing and orders to build a car.

His story, in the material particular of the manner in which he obtained possession of the drawing, is improbable on its face, and strongly contradicted by the testimony in the case.

At page 45, plaintiff's proofs, folio 208, *Cromwell says:* " *The drawing shows the body and the running gear, as the carriage was built.*"

In plaintiff's Troy case, page 67. Reuben Aler testifies: " It differs in the trucks; the said drawing shows the perch to run all the way through, while, to the best of my recollection, the perch only ran half way through in the Columbus. My recollection is, that the perch stopped at the bolster, and was connected with a pair of hownes. The panels of the body in said drawing appear to me to be wider than they were in the Columbus I think the railing on top is represented differently, but I cannot say positively."

Cromwell is also contradicted on that point by Jacob Rupp, defendants' No. 4, pages 22, 23—folios 84 to 87; by Henry Shultz, defendants' No. 4, page 41—folios 155 to 157; by Edward Gillingham, defendants' No. 4, page 57—folios 220, 221; by Conduce Gatch, defendants' No. 4, pages 25, 26—folios 94 to 101; by Leonard Forrest, defendants' No. 4, pages 37, 38—folios 140 to 146; by Jno. M. Eichelberger, defendants' No. 4, pages 20, 21—folios 75 to 80, and defendants' No. 4, pages 36, 37—folios 135 to 139; by James B. Dorsey, defendants' No. 4, pages 76, 77—folios 289 to 290; by the " Baltimore Patriot," referred to in the affidavit of John F. McJilton, defendants' No. 4, page 39—folios 148, 149, of date June 24th, 1831. Plaintiff's proofs, pages 112, 114, vouchers Nos. 31 and 39, also show that Conduce Gatch was superintendent, or master carpenter; that the work was done under his direction, and that he received moneys and paid all the hands in the department.

The counsel on the opposite side attempted to make it appear that Gatch had said at one time that Winans gave him the drawing, and, at another, that it was given to him by George Brown. Gatch's testimony here shows that the drawing was *furnished* by Winans, and *handed to him* by George

Brown, which reconciles this apparent discrepancy. Cromwell's story, that Winans gave him the drawing, in folio 221, page 48, plaintiff's proofs, therefore, is quite improbable, as he was not the principal; and being a journeyman, not the person to whom it would be given. But Gatch was the proper man to receive it, as he was appointed, and had power to direct and superintend the work to be done in relation to the cars. Cromwell's idea is very extraordinary—that the mere lengthening of a car shop, as he says, should impress upon his mind the circumstances connected with the drawing, as he states them on page 46, folio 211, plaintiff's proofs. He evidently is mistaken; his story is erroneous, it is a delusion.

At page 30, plaintiff's Troy case, Cromwell says: "I had the drawing in my possession for seven years. I put it up over the stove hole;" which, it is alleged, was the cause of the drawing being *smoked irregularly*, as it appears.

Mr. John A. Currie states, defendants' proof No. 4, page 93: "*Smoking drawings or paintings on paper is the mode generally resorted to, to give new paintings an old appearance.*"

The drawing does not represent the Columbus; is not the same from which the car was built; was not handed to Gatch, nor seen by our witnesses. Cromwell was a journeyman, and produces it, *smoked*, at the Newcastle and Frenchtown trial.

Taking all these circumstances into consideration, it appears that the smoked drawing, *irregularly* smoked as it is, was manufactured for the Newcastle and Frenchtown case; and that it was thus smoked to conceal the fact that it had not been used in building the car. If it had been evenly smoked on the face, the fact that it had not been soiled, and not been used, would have been apparent from its fresh and new, or even and unsoiled, aprearance; but it is smoked in shades, or irregular—as it would have been soiled by use—to conceal the fact that it had not been used.

It further appears, that Cromwell has been acting as agent of the plaintiff, from the testimony of Jacob Rupp, defendants' No. 4, page 23—folio 87, where he says: "Last fall, a considerable time after I had given my affidavit in this case,

Mr. Oliver Cromwell and Mr. Gould, and some person with them who I do not know, called on me at my residence in St. Mary's-street, Baltimore. I was sick at the time; and they insisted on talking to me about this matter, and the affidavit I had given. Mr. Gould said he was astonished at my affidavit; and said that he had five companies to fight, and that I ought to lean on his side just a little."

Michael Glenn, it appears, also called on John P. Mittan, as Mr. Mittan's affidavit shows, page 60, defendants' No. 4—folio 229. He says: "About ten or twelve days ago, Michael Glenn brought a person to me and introduced him; he took my affidavit."

Then, Oliver Cromwell and Michael Glenn, the plaintiff's two principal witnesses, it appears, were *both agents of the plaintiff*, according to the testimony of Jacob Rupp and Jno. P. Mittan.

John M. Eichelberger, at page 20, defendants' No. 4—folio 77, states further: "Oliver Cromwell, and two persons with him whom I did not know, called on me at the corner of Paca and German streets, Baltimore."

All of these facts and circumstances certainly show that no reliance ought to be placed on Oliver Cromwell.

Henry R. Reynolds is the next witness for the plaintiff. We will prove that his statement is erroneous. He says, as to the carriage and horse car of December, 1830, at page 52—folio 241: "That one of the four-wheel cars, used for wood was taken, and boards nailed on the standards. The horses were then placed on the said car, and other boards were nailed across the end, so as to inclose the horses on said car, and prevent them from being injured while passing over the road. Another car was then obtained, on which the carriage was placed; and the two cars—one with the horses as a load, and the other with the carriage as another load—were sent from the said Mount Clare depot up the said road."

Now we submit, as a common sense conclusion, that it is not at all probable that simply nailing boards on the standards of cars out on the road would have been entered as "two trussels for the transportation of horses and carriages," and returned as work done in the shops.

In plaintiff's proofs, page 109, we find entered, as work done in the shop, in December, 1830, "two trussels for the transportation of horses and carriages."

At page 104, plaintiff's proofs, in answer to the question, "Will you look at the papers now shown to you, marked 'A, B, C, D and E,' and state what they are?" Mr. Gatch says: "They are monthly pay-rolls; vouchers for materials purchased, *and reports of work done in the month in the shop,* for the respective months specified in them."

In reply to the next question, on page 104, plaintiff's proofs, Mr. Gatch says: "The hands were sometimes employed for other purposes than building cars. Such work was generally charged to the department it belonged to, so as to charge it to the construction of the road, or wherever it belonged. *The monthly reports did not mention work done about the yard, such as repairing cars and such like; it was of work done and finished in the shop. There was no other report made by me.*"

That shows conclusively that the nailing of these boards on the standards was not work done in the shop, and not making trussels, and not such work as was reported. In connection with this, I will read a publication to be found in the affidavit of Thomas Murphy, defendants' No. 4, p. 55—fols. 210 to 213.

Mr. Murphy states: "I am seventy-one years of age, and have been one of the proprietors and publishers of the newspaper printed and published in the city of Baltimore, entitled the 'American and Commercial Daily Advertiser,' from the 1st July, 1810, up to the 30th June, 1853; and have kept regular office files of the said newspaper, in the office thereof in the city of Baltimore, during the said period, to which files I now have access. In the said newspaper, printed and published on December 18, 1830, are the following words— The heading reads 'American and Commercial Daily Advertiser. Vol. LXII. Baltimore, Saturday morning, December 18,

1830. Whole No. 9871.' And at the head of the second editorial column of said paper of December 18, 1830, is the following article: 'We learn from the Gazette, that ex-President Adams and his lady, Sir William Campbell, and several other strangers, accompanied by the president and some of the directors of the Baltimore and Ohio Railroad Company, made an excursion up the road yesterday morning. *Mrs. Adams' carriage, with the horses, was, we learn, placed upon a trussel attached to one of the railroad cars,* and taken as far as to the Relay House, at Elk Ridge Landing, from which place it proceeded to Washington by the turnpike road.

"' *We understand that the railroad company have prepared several of these trussels, by which either wagons or pleasure carriages may be transported along the road without being unloaded or putting the passengers to the inconvenience of getting out of their own carriages. This is certainly a new and important facility which this admirable system is capable of affording.*'

" The above is a true copy of the whole of the said publication, and it was printed and published in the said newspaper on the said 18th of December, 1830, and I believe the facts therein stated to be correct; I never heard them contradicted nor their truth questioned."

That this was an eight-wheel car is also proved by Edward Gillingham, by Jacob Rupp, and in the plaintiff's suppressed affidavit, presented by the defendants; by Conduce Gatch and by Leonard Forrest, who, at that time, were higher in position, and older and more responsible men, than Glenn and Reynolds. He and Reynolds must refer to other circumstances and cars, entirely, if any such boards were nailed up as he states.

Thomas Walmsley, plaintiff's witness, I will refer to as unreliable. In plaintiff's proofs, page 54—folio 248, says: " That he has the impression, but is not positive, that the said four-wheeled cars were always used singly."

He declined, therefore, to swear positively that there were no eight-wheeled wood and trussel cars.

I will here remark, generally, that the plaintiff's witnesses show that they had no idea of the mechanical organization of an eight-wheel wood car or trussel car, as consisting of two ordinary four-wheel trucks, with king-bolts and bolsters supporting the body. And, not having this idea in their minds, they could not negative that construction and arrangement, and hence their affidavits are, throughout, ambiguous.

Mr. Walmsley, at folio 251—page 54 of plaintiff's proofs, speaks of the drawing, and professes to recollect more, after a lapse of twenty-two years, than he can certify to with the matter immediately before him. In fact, all the plaintiff's witnesses swear from impressions or belief as to the drawings; while, with drawings before them, and others presented to them as copies, they cannot determine whether they are true copies of the drawings or not; they testify that the copies are true, *omitting a vital part*, the king-bolts, to which I have before referred. And yet they say, on their oaths, that they believe the smoked drawing to be the original. We believe that it is manifestly easier to compare two drawings, than recollect one for twenty-two years.

The plaintiff also took an affidavit of Jno. P. Mittan; we took a cross-affidavit. In the plaintiff's affidavit, Mr. Mittan is represented as using the same indefinite language, as to mechanics, as the others; he, however, does not say that Winans was the inventor of the eight-wheel car. In the plaintiff's affidavit he is represented as being employed in building burden cars; whereas, we show that he was engaged *in planing up green wood alone*. He did not know the principle and construction of the eight-wheel car. He says he did not notice the cars; and hence his affidavit, as produced by plaintiff, was written in the same strain with the others, as though each one recollected all that was done and going on; when the fact is, that they were not in a situation and capacitated to observe, particularly and exactly, the mechanical principle and construction of the cars, and consequently were not competent to testify on the subject.

In the plaintiff's version of Mr. Mittan's affidavit, it is stated that the Columbus drew by the body, but the time when it did so is entirely omitted, thus endeavoring to make appear as if it were soon after the Columbus was built; whereas, Mr. Mittan, as shown by our affidavit, was away from Baltimore, out of the employ of the Baltimore and Ohio Railroad Company, six years; and the time of which he speaks is in 1837, when he returned and saw the Columbus in an altered condition.

I will now refer to the testimony of Mr. Jeremiah Myers, plaintiff's proofs, page 85—folios 400, 401. The plaintiff pretends that the *first eight-wheel cars used on any roads going out of Boston* were built by him at Attleboro, for a branch of the Boston and Lowell railroad, by measurement made by him personally at Philadelphia, from an eight-wheel car built at Baltimore upon the plan and construction then USED by the patentee, and that the plan was the same as described and laid down in the patent.

This affidavit, it appears to us, is intended to give a false impression as to the *time* when Myers went to Philadelphia, and a false statement of the *facts*. The time is not stated, neither the road where he made the measurement. He also conceals the measurements; and not only so, but the construction, or machinery, is not stated or described.

The time at which Myers went to Philadelphia was not till 1837, since no eight-wheel cars were introduced on the Boston and Lowell road until this time, as shown by the affidavit of Mr. Lee, page 112—folios 445, 446 of defendants' No. 3; and Mr. Grigg's, defendants' No. 3, page 105—folios 417, 418.

The statement of Mr. Myers is also false in substance, because not one car was constructed according to Winans' plan, as shown by Mr. Lee, defendants' No. 3, page 112—folios 445, 446; and Mr. Higginson, engineer of the road, defendants' No. 4, page 83—folio 314; and the Boston and Lowell road only began to operate in 1835, as shown by Mr. Higginson's affidavit, same reference.

Mr. Myers made entirely satisfactory eight-wheel cars, by joining the bodies of two four-wheel cars on long timbers.

Now, he pretends that he used the old wheels, and made trucks according to Winans' patent, and placed them on the Boston and Providence road, and that they were there used for many years.

The affidavit of Mr. Griggs, in defendants' No. 3, page 104—folio 414, shows that the first eight-wheel cars on the Boston and Providence road—to which Myers also refers—were built in 1838; that the distance of the bearing points of the wheels apart was *five* feet; that the breadth of the track was four feet eight and a half inches; and that the distance between the wheels was twenty-seven inches. So that it is not true that those cars, nor any cars, were ever built according to Winans' patent. Myers does not state that he ever built a car *from the specification*.

Mr. Griggs, defendants' No. 3, page 110—folio 438, says that the wheels in all the bearing carriages are about the same distance apart, and that the frames are rigid and square.

Myers says that he measured a *Philadelphia car*, and built them according to *said measurements;* and the fact that the cars on the Boston and Lowell, and Boston and Providence roads were constructed with the machinery that Mr. Griggs and Mr. Lee describe and set forth, proves conclusively that they were not in accordance with Winans' patent. The bearing points of the wheels were as far, and in some cases further apart than the guage of the track. The trucks were rigid wheel-frames, and about seven feet from the ends of the body.

I will also refer, upon that point, to the affidavits of Mr. Parrott, defendants' No. 3, page 152—folios 608 and 612; and Godfrey B. King, defendants' No. 4, page 79—folios 298, 299, 309; page 81—folios 308, 309. We say that, after all, it appears by the testimony that instead of adopting Winans' running gear, his plan, or his theory, they adopted the Quincy car, so far as the running gear was concerned, as to the distance of axles, and the stiff, rigid wheel frame truck and side bearings; and the body was, of course, fashioned for convenience. As Myers does not give the measurements that he took, it cannot be inferred that he measured the trucks or running gear; and, in fact, he does not pretend to have done so. At the time Myers went to Philadelphia, and measured a car there, eight-wheel passenger cars were in use all over the country, viz: in 1838. He went there two years after the car "Victory" had sprung into existence, and after many other cars had been built from her.

The intent of his statement is to create an impression that the car he measured came from Ross Winans. This is also contradicted by Henry Shultz, car builder, in defendants' No. 4, page 40—folio 154; and pages 43, 44—folios 165 to 170, who states that no car whatever, constructed according to Winans' specification, was in use on the Philadelphia, Wilmington and Baltimore railroad.

I will now speak of a matter in which my very highly respected friend, Mr. Latrobe, who was a witness in the Troy case, is involved; and which relates to this drawing of the Dromedary, now before the court, a copy of which is annexed as an exhibit to the testimony of Michael M. Glenn, page 40 plaintiff's proofs, folio 182, and page 40—folio 190. Of Oliver Cromwell, pages 47, 48—fols. 218 to 221; of John Ferry, page 59—folio 274; and Thomas Davis, page 61—folios 283, 285.

To contradict the statement of these witnesses, which is, that this drawing represents the Dromedary, I will refer to the testimony of John H. B. Latrobe, defendants' Troy case, pages 19, 20, and plaintiff's Troy case, page 23, in his answer to the 3d interrogatory:

Mr. LATROBE.—I have not the slightest confidence in my own memory.

Mr. HUBBELL.—At the same time, as you are contradicting your own witnesses, and as you set them up as perfectly reliable to contradict ours, I am bound to show that in which you differ from them, that you do not agree among yourselves. To the third interrogatory, Mr. Latrobe answers: "That he recollects distinctly three cars of eight wheels built after the Columbus, and put on the railroad in 1834. The first was named the Winchester, which was composed of three carriage bodies, built with swelled sides, after the fashion of common coaches, and resting on a framing, which in its turn rested on two trucks, which, as near as witness recollects, were under the centre of the outer bodies. The next car was called the 'Comet,' and consisted of five small carriage bodies, the three inner ones being suspended between the trucks, and the two outer ones being immediately over the trucks.

" These bodies rested on a platform, composed of two side-pieces, plated with iron, and curved so as to permit the inner bodies to hang between; the trucks were composed of four wheels each; *and the bolster, instead of resting upon an ordinary frame, rested upon strong springs, that connected the axles as they rested upon the outside journals. The wheels of the trucks were brought as close together as could be,* leaving room for the lower bolster to rest on the springs between them. The trucks were under the centres of the outer bodies. This car was subsequently altered, but the date is not recollected by witness, so as to put all the bodies on the same line. Witness appends hereto a rough diagram, marked 'J. H. B. L., No. 1,' to explain his testimony more fully as regards the form and construction of the cars above mentioned.

"The last car was the 'Dromedary,' which witness believes to be represented by the drawing, marked 'J. H. B. L., No. 2,' *with the exception that the spaces over the trucks were closed up, so as to make a sort of box or car body.*

" Deponent does not recollect when these cars were commenced; he speaks of the order only in which they appeared on the road; he believes they were all being built at the same time." All the other plaintiff's witnesses swear without reservation that the drawing represents the Dromedary.

This shows that the witnesses do not fully understand the drawings and arrangements about which they speak; for the drawing, it is plain, has not the body extending over the trucks the car itself had. Their descriptions are inaccurate in regard to highly important matters in this case, with reference to the drawings, and are not reliable.

I had intended to call the attention of the court to the kind of wheel used on railroads, and also to the manner of applying springs and pedestals, as used by the defendants; but for the present I will, in regard to the wheel, refer to Tredgold's Treatise, published in 1825, pages 42, 43, fig. 24, plate 3, where it is shown:

Page 42–3. "*Resistance of the surfaces of the rails.*—It often happens that the great part of the resistance of the rails arises from the lateral rubbing of the guides of the wheels; therefore, it is desirable to give the wheels a tendency to keep in their path with as little assistance from the guides as possible. For edge-rail carriages, this may be accomplished by making *the rims of the wheels slightly conical, or rather curved, as shown in fig.* 24. The carriage will then return of itself to its proper position on the rails, if it be disturbed from it by any irregularity." Mr. Hubbell here explains the figure, and states that this is the kind of wheel now in general use by the defendants and railroads, the conical tread and flange being in all cases alike; the connection between the hub and the rim being by spokes in some; in others, by compensating plates, shaped in various ways.

For the mode of applying springs to the truck-frames, we refer to plate 111, chap. iv., pages 75, 76, Wood's Treatise of 1831; and to the Novelty, in Earle's Treatise of 1830.

In Wood's, the pedestal, holding the box of the journal, and confining the action of the spring normal or perpendicular to the track, is fully described and shown. That is the plan now in general use on both four-wheel and eight-wheel cars.

My next point is, as to the extent of railroad track laid in May, 1830, east and west of Mount Clare. That the extent of railroad track laid in May, 1830, west of Mount Clare, was about 13 miles; and the amount laid east of Mount Clare, into the suburbs of Baltimore, was about one mile. I make this reference in order to rebut plaintiff's assertion, that Rutter and May—our witnesses—were in error as to the track laid in 1830 and 1831 and '32.

That the road was open on May 22d, 1830, to Ellicott's Mills, a distance of about 13 miles, I refer to the testimony of Jno. Elgar, plaintiff's proofs, page 24—folio 100. As to the portion of road opened east of Mount Clare into the suburbs of Baltimore, I refer to the 1st volume and 4th annual report of the Baltimore and Ohio Railroad Company, page 91.

In the report of James P. Stabler, assistant engineer, dated Sept. 30, 1830, addressed to Jonathan Knight, chief engineer, it is stated: "It will be remembered that, previous to the determination to run the flanges on the inside of the rails, a single track had been laid *on the city division*, to accommodate the arrangement for the flange on the outside. This change in the mode of using the flange required an alteration in the said track, *which was made by contract during the months of April and May.* The expense attending this alteration was as follows, viz.:

"For the straight line, *east of Mount Clare*, $1.75; and for the remainder $2.50, per rod run.

Signed, "JAMES P. STABLER,
"*Assistant Enginçer.*"

In the same report, dated October 1st, 1830, we find the following:

" *The City Division.*—*This division, which commences at Pratt-street, extends to the 'first stone,' or western boundary of the city, and embraces a distance of one mile and* $111\frac{32}{100}$ *poles,* was not put under contract until about the middle of June, 1829."

Signed by
" CASPAR W. WEAVER,
" *Superintendent.*"

The road, therefore, *was laid east of Mount Clare, about one mile, in April and May*, 1830; and the statement of Rutter, that he saw the eight-wheel wood cars on it in the fall and winter of 1830, stands sustained; and of Edward May, that the rails were laid up Charles-street, to connect with it in the early part of 1832, is also sustained.

Without further comment on this point, I proceed to my next subject, which is Mr. Gatch's position.

CONDUCE GATCH.

He is one of the most important witnesses on behalf of the defendants, and has been very violently attacked by the opening counsel for the plaintiff, and my object is to show, as briefly as possible, his position.

In the defendants' Troy case, page 33, it appears that he is a *millwright;* that is his trade or occupation, and was at the time of his employment by the Baltimore and Ohio Railroad Company. He belonged to a class of men like Oliver Evans, the inventor of the locomotive steam engine, and numerous other improvements in various branches of machinery, who have been among the foremost in improvements, and who, although millwrights, have not been excelled in ability and invention in any branch of the mechanic arts by any other class of mechanics; and are as fully competent to invent, even more competent, than the majority of other men. Experience has singularly shown this to be the fact.

In defendants' Troy case, at page 35, he testifies that " he had the direction of car building." And also in plaintiff's proofs, page 103—folios 490, 491; and page 104—folios 495, 496, he says: " That he had direction of the car building; he employed and controlled the hands; received and paid them their wages, and conducted the whole business belonging to that department." In this he is also fully corroborated by the vouchers produced by the plaintiff himself, on pages 106 to 114, inclusive, of plaintiff's proofs. I will read from one of them, on pages 108, 109:

ARGUMENT OF

"(C. GATCH.)   (VOUCHER No. 58.)   (B.)

"We, the undersigned, do acknowledge to have received of C. Gatch, master carpenter, the amount hereunto set opposite our names, respectively, in full payment of our services for the time herein specified, while employed under his direction on duties relating to the construction of Wagons, &c., for the Baltimore and Ohio Railroad Company. December, 1830.

No.	NAMES.	Occupation or nature of expenditure.	No. of voucher.	Term of service.	No. of days.	Rate per day.	Aggregate.	Signers' names.	Witnesses' names, &c.
1	C. Gatch,	Master Carpenter,	....	From Dec. 1, to Dec. 31.	26	$2 00	$52 00	C. Gatch,	
2	M. Glenn,	Carpenter,	....	do	13	1 25	16 25	M. Glenn,	
3	J. Rupp,	do	....	do	26	1 25	32 50	J. × Rupp,	F. A. Gatch.
4	F. A. Gatch,	do	....	do	16	1 25	20 00	F. A. Gatch,	
5	W. O. Frost,	do	....	do	26	1 25	32 50	W. O. Frost,	
6	E. Eichelberger,	do	....	do	26	1 25	32 50	E. Eichelberger,	
7	J. Eichelberger,	Painter,	....	do	16	1 50	24 00	J. Eichelberger,	
8	O. Cromwell,	Coachmaker,	....	do	26	1 50	39 00	O. Cromwell,	
9	H. Reynolds,	Carpenter,	....	do	26	1 25	32 50	Henry Reynolds	
10	L. Forrest,	Smith,	....	do	26	1 42	26 92	L. × Forrest,	O. Cromwell
11	C. Parraway,	Helper,	....	do	26	75	19 50	C. × Parraway	
12	P. Fullerton,	Laborer,	....	do	26	75	19 50	P. Fullerton,	
13	D. McAvoy,	Drayman,	1	....	....	....	6 91½		
14	R. Greenough,	Stone, window, sash & glass,	2	....	....	....	4 00		
15	Deaver, Tibles & Jenkins,	Turning leather nuts, &c. &c.	3	....	....	....	7 75		
16	Wm. S. Browning,	Hardware, glass, oil & paints,	4	....	....	....	19 16		
17	Gillingham & Jessup,	Hardware,	5	....	....	....	21 91		
							$416 90½		

"I certify that the above account, amounting to $416.90½ is correct, and that the articles and items therein charged were procured and applied in aid of the construction of the Wagon for the Baltimore and Ohio Railroad Company, December 31, 1830.   C. GATCH, M. C.

Amount, ... $416 90½
Error in No. 1, ...   10

$416 80¾

"Baltimore, 1st January, 1831, Received of Geo. Brown, Treasurer, the amount of the above account.

C. GATCH.

"I believe the above is correct.

"Audited. Four hundred and sixteen dollars and eighty and a half cents, (416.80½.)

"(Approved.) P. E. THOMAS, P.

G. BROWN, *Treasurer.*
W. WOODVILLE, *Auditor.*
JOHN N. BROWN.

Work done in December, 1830.

" Two plain box wagons, Nos. 41 and 42.
" *Two trussels for the transportation of horses and carriages.*
" One baggage wagon.
" One plain box wagon, No. 43.
" *To fitting up Col. Long's wagon with springs, glass, &c.*
" To making 160 lights sash for shop.
" To making 4 plain box wagons, Nos. 44, 45, 46 and 47.
" To making new sides for coach, with brakes, complete.
" 1 new pair of shafts; 13 pair repaired.
" Smith work for the above; painting."

He then, as shown, was a person in the trust and full confidence of the Baltimore and Ohio Railroad Company, and was not a mere hired workman, as it has been insinuated, for the purpose of casting odium upon him. No greater reliance could have been placed in any man than was reposed in Mr. Gatch; and therefore the imputations attempted to be fastened on him are not only disproved by the vouchers produced by plaintiff, but display an improper and prejudiced feeling on the part of the plaintiff's counsel and witnesses. And at this time, also, there is no more respectable man, nor any more deserving of confidence and esteem, than Conduce Gatch. He is a mechanic of the highest order of intellect; sound and unflinching in his views; of inflexible integrity; and all the stigmas that have been hurled at him fall harmless at the feet of such a man; they rather recoil on those who have aimed them at him.

With a view further to show that Mr. Gatch took charge of the shops, and employed men to work under his direction, as early as the 3d April, 1830, I refer to his affidavit in defendants' No. 4, page 35—folios 131, 132. And I introduce this here for the double purpose of showing the *time* at which the persons there named commenced their work under his direction, and that *he was in full charge of this department as early as April 3d*, 1830.

Timber Cars, 1830.

The next point to which I shall allude is as to the *timber cars*. And here I make this proposition: That, under Chief Justice Taney's instructions, the timber cars, on eight wheels, used in the spring of 1830, embodied the same principle and mechanical construction as the cars of the defendants. I say again:

1st. That they practically embodied the principle of the eight-wheel car, as built by the defendants, in the fact that the timbers, as secured by the standards, held by their own weight on the bolsters, and the trucks swivelled under the body, burden, or load.

2d. No new principle, nor substantially new arrangement would be effected by bolting the body of timber fast to the bolsters, to make the connection more permanent. Such an act would be no new invention.

The plaintiff, I understand, admits that the timber cars existed before Winans' alleged invention; and therefore I will not refer to all the testimony on this matter.

I will, however, call the attention of the court to the affidavit of Conduce Gatch, in defendants' No. 1, page 29—folios 181, 186, and particularly to folio 183, as to the distance of the *axles* apart, which he states was "two feet ten inches in each of the trucks." Also, the deposition of Jacob Rupp, defendants' No. 2, pages 14, 15—folios 53 to 58. Also, to the testimony of Thomas McMechen, in defendants' No. 3, pages 41, 42, 43—folios 163 to 170, inclusive. Jacob Rupp there testifies that "he built these cars by Gatch's instructions, and under his directions;" that "they were used in May, 1830;" that "the distance, apart, of the wheels in each truck was about six or eight inches;" and that "the timber held by its own weight;" "and that persons often rode on the timber."

The counsel for the plaintiff has contended that the timber will slip; and in order to illustrate this, a piece of wood, proportionately a little larger than and in the shape of a lath, that a school-boy could carry, was brought into court.

When the cars are loaded, they have a very heavy weight or load of timber upon them, causing, of course, considerable adhesion, or violent contact between the bolsters and the body of the timber.

Now, till the counsel can prove—which he never can—that it requires less power to cause timber to slip or slide on bolsters than it does to compel the wheels of a carriage to revolve on their diminished journals, he cannot demonstrate that the timber would slip. It is perfectly manifest that the weight of the timber will hold with greater tenacity to the bolsters than will the lubricated axles to the boxes. It requires the one four-hundreth part of the weight only to draw a car, by causing the wheels to revolve; and that timber would remain in a body on the bolsters, just as steady as though it had fifty spikes driven through it into the bolsters. The simple addition of spikes driven through the timber does not enable the trucks to fulfill their functions with greater completeness, neither does it alter the principle of the car and the physical organization of the running gear. And when a man comes before a court of justice, and attempts to convice it that more power is required to cause the wheels to move on their axles, lubricated as they are, than to compel the timbers to slip, he undertakes to prove that which no living man can prove. Were such the case, wheels had better be dispensed with altogether, and freight and passengers be propelled on a sledge, to slip or slide along the rails without the use of wheels.

Such arguments as plaintiff's are very frail and feeble. They will not answer for a sound, practical mind; and hence Judge Taney's views are to the point precisely. We, however, do not use the timber cars to destroy the plaintiff's patent, unless it should be construed so largely as to include both the defendants' cars and the timber cars.

In the fourth section of Chief Justice Taney's instructions to the jury, that judge shows that he knew what mechanical principles were, and he knew the relations between a rolling and sliding surface; and the very basis and elements of mechanics is infused into those views and instructions.

The chief justice says, at section 4, page 21, plaintiff's proofs: "The plaintiff having claimed, in his specification, as his invention, *the manner of arranging and connecting the eight wheels of a railroad carriage, in the manner mentioned in his patent,* for the purpose of enabling burthen and passenger cars to pursue a more smooth, even and safe course over the curves and irregularities of a railroad, *he is not entitled to recover*, if the jury find from the evidence that, before the time when the plaintiff claims to have made this discovery, carriages with eight wheels, arranged and connected *substantially in the same manner, and upon the same substantial principles* with those described by the plaintiff in his patent, were known and had been publicly used *for the purpose of transporting long timber more smoothly, evenly and safely over the curves and irregularities of a railroad than could have been done in cars of the ordinary construction, with four wheels. And the circumstance that those timber cars were used only for temporary purposes, and were formed by using two ordinary four-wheel cars, with the hind and for wheels close together, as bearing carriages, and without any car body upon them, cannot affect the question now before the court.*"

These latter words show that the judge understood mechanics sufficiently well to know that the draft applied to cause the wheels to move did not affect the whole body of timber, in its position on the bolsters, in relation to the active principle of the arrangement.

The defendants in that case, however, asked that such organization should be considered as invalidating the plaintiff's patent. But, under our construction and view of Winans' patent, we do not present it in that light here. We present it as an organization of the principle of an eight-wheel car, or swivelling motion of two four-wheel trucks, under the one body, burden, or load, with the trucks constructed, as defendants' witnesses testify, of rigid wheel-frames, and the wheels six or eight inches apart in each truck; substantially the same construction and action as the defendants' cars, with the exception that the wheels are *closer together* than they manufacture them, although they are not so close as Winans' specification calls for.

One of these timber cars, as shown by Thomas McMechen, "was permanently connected," to distinguish it from permanently holding by the load. It was connected by a piece extending from the king-bolt of one truck to the king-bolt of the other.

To explain what he states, more fully, I will read his evidence on this subject:

He says: "I was one of the superintendents and contractors on the Baltimore and Ohio railroad, in the years eighteen hundred and twenty-nine and eighteen hundred and thirty. I was superintendent in 1829, and contractor in 1830.

While I was contractor, in 1830, between the month of March and the month of May, the long timbers, called string-pieces, used in making the track — the pieces being six inches square, and from sixteen to forty-eight feet in length — were hauled on the track nearest Baltimore, on eight-wheeled cars; and, by means of these eight-wheeled cars, they were delivered wherever required. One of these eight-wheeled cars consisted of two four-wheel trucks, having regular truck-frames, holding the axles of the pairs of wheels parallel to each other, and the wheels in each truck were about six or eight inches apart; and in this space between them was a brake, similar to what is in use now on various roads in eight-wheeled cars. The frames of these trucks had stout cross pieces for bolsters, and were planked over with the other part of the framing, holding the whole frame firmly together. On top of each of these four-wheel trucks was placed an upper bolster, secured to the middle of the trucks by a vertical centre pivot; and these two trucks, with their bolsters, were permanently fastened together, a considerable distance apart — about twenty feet apart — by a long string-piece, six inches square; being one of the pieces, the same as the rails were laid with. This piece of timber secured the trucks this distance apart by resting on the middle of the bolsters, and the centre pivot of the bolsters going down through a hole in each end of the timber, and through the bolsters.

"Upon top of the bolsters, also, of this eight-wheeled carriage, thus permanently connected or organized, the long timbers or string-pieces were placed and transported on the road wherever required; and on their removal from the car, the car still remained as before, an organized body, and was drawn back for another load. This was the construction and operation of the eight-wheeled car, that delivered the timber on the road along side of my contract.

"The car ran on a road that was laid in 1829, and on the track that it was running or finishing out; and I saw the car daily. I finished my contract in May, 1830, and, after visiting Pottsville, Pennsylvania, I returned; and in July, 1830, for about three weeks, I superintended the laying of rails on the second division of the Baltimore and Ohio railroad, beyond or west of Ellicott's Mills; and after this, in August, 1830, I was attached to the engineer corps, on the survey of the Baltimore and Susquehanna railroad, and had no further connection with the Baltimore and Ohio railroad. I am certain that this eight-wheeled car, constructed and operated as I have described, was used during the latter part of April and fore part of May, 1830; and persons used to often ride on it when it was running on the road. The trucks turned under the bolsters, to suit the curves and other inequalities of the road; and in point of principle of construction and operation, my opinion is, that it was the same substantially as the eight-wheeled platform, and freight and passenger cars, now used on the various roads."

The connection with a bolt or pivot through the bolster, was an arrangement which could be accomplished in 15 minutes; and the only difference between that and the timber resting on its own weight is, that the one has the bolt through it, and the other has not; but depends on the adhesion of its great weight alone. The timber connected with its own weight and side standards, was just as secure for the practical purpose of transportation and use of the swivelling principle. It developed the combination and organization for the time, as much as though it had been bolted there; and, as Judge Taney very properly says: " The circumstance that those timber cars were used only for temporary purposes, &c.," or "*for a limited time, without any car body on them*" (for all cars and machines are *limited* when it comes to time), "*cannot affect the question now before the court.*"

It certainly cannot be contended, for a moment, that the exercise of a common mechanical act in driving spikes through boards is the invention of an eight-wheel car; and yet, mechanically, it comes down to that. The principle, the operation, the user, are there proved in 1830, and that is everything, as Chief Justice Taney's instructions show; for the same impression is so apparent in that charge that it cannot be viewed in any other light; and we say that they are substantially—in their features of construction and action, in their principle and general character—the same as the cars made by the defendants, as distinguished from the plaintiff's specification.

This is also confirmed by Leonard Forrest, in defendants' No. 2, page 18—folio 74, who says: " That the timber cars were the same as the wood cars, with the exception that the timber laid by its own weight," and thus formed the connection and organization in principle, the same as defendants use.

Wood Cars, 1830.

My next object will be to prove that eight-wheel, double-truck, swivelling, long-bodied wood cars, for the transportation of amounts of cord-wood, were made and put in use, in the ordinary business of transportation on the Baltimore and Ohio railroad, as early as November, 1830, and continued in use for years; thus being prior to the Columbus, and embodying all the essential principles of that car, so far as relates to the running gear, and the connection of the same with the frame or body.

Mr. Elgar's testimony as to the time when the smoked drawing of the Columbus was produced, and the facts connected with it, was contradicted by himself in the Troy case, and by all the other witnesses of the plaintiff. The testimony on both shows that there was a drawing in February, 1831; whilst these eight-wheel wood cars went into operation in November, 1830, prior to the alleged drawing some three or four months. To prove this, we rely upon the following testimony:

First.—Conduce Gatch, defendants' No. 2, pages 20, 21—folios 81 to 84; No. 4, page 24—folios 90 to 94; No. 4, pages 34, 35—folios 129 to 134.

The construction of these wood cars, as differing from the timber cars, consisted in driving bolts through the string-pieces, which had been procured for the construction of the road; a slight change, requiring but a few hours' work. Mr. Gatch testifies that these cars *were fitted up on the track* in that manner, *with string-pieces or timbers originally procured for the construction of the road, in November,* 1830, *and that they were used for the transportation of cord-wood.* We see at once that the first article, naturally presented for transportation on a railroad, would be the wood in the line of the road; and we also see how natural it would be, after having used these timber cars, organized in May, 1830, and during the following months, for the construction of the road—how natural it would be for any mechanic, having charge of the construction of cars, to drive bolts through two of these string-pieces for the purpose of laying the cord-wood thereon, and to move the standards to support the wood at the ends.

To the testimony of Jacob Rupp, defendants' No. 2, pages 14, 15—folios 54 to 59, and 65; defendants' No. 4, page 22 —folios 82 to 84; defendants' No. 4, page 62—folio 239.

Mr. Rupp assisted in making these cars. He was a prominent workman; and a more open, honorable, honest man than Jacob Rupp is not to be found in the city of Baltimore. He has, by his uprightness and industry, made himself comfortably well off. Such a man is an ornament to the community in which he lives, and to a country like this. Upon such men the country is to rely for its stability and practical advancement; and such men are entitled to great weight and consideration in a court of justice. An attempt has been made to stigmatize him as a *workman,* as though there were an aristocracy and a feudal system here; as if some were to be believed because they wore fine clothes or assumed official distinctions, and others not, because they were industrious, and earned their bread by the sweat of their brow. We know no such doctrines and principles in this republic; and we depart from the true character of our government, and the fundamental principles of our institutions, when such notions as this are sustained, or become popular.

By the testimony of Leonard Forrest, defendants' No. 2, pages 18, 19—folios 72 to 77; defendants' No. 4, page 17—

folios 62 to 63; defendants' No. 4, page 18—folio 67; defendants' No. 4, page 19—folio 73; defendants' No. 4, page 38—folio 146.

Mr. Forrest is a man with none to excel him in his branch of business. The very best of work produced by smiths is the work of Leonard Forrest; and it is so known, generally, in the city of Baltimore. A more reliable man—a more sound man—a more skilfull man in the working of iron—a man of more strength of mind, tenacious memory and accurate observation—is not to be found than Mr. Forrest. He testifies to the same mode of building these wood cars; and, as he did the iron work for them, it cannot be supposed that he is ignorant of the fact, and the circumstances connected therewith.

Better witnesses than these cannot be obtained. To whom does the law look for knowledge and evidence on such subjects? Will this court look to men who were engaged in keeping books, without sufficient mechanical discrimination to distinguish between one piece of machinery and another, as to its arrangement, construction and development of action, rather than to those who did the work, who had it before them—not only by observation, but who had it in their minds to give it form—to make it—to create it? Certainly not. The latter is the class, and whom we have produced; to whom the law looks, and to whom the law would send us, for the purpose of ascertaining the facts.

By the testimony of John A. McClarie, defendants' No. 3, pages 83, 84—folios 331 to 334; and Murphy's affidavit, immediately following.

McClarie is now in the prime of life, and at that time was a smart and observing boy; and there is this fact connected with him—that his pursuits in after life show—that his tendency of mind was towards railroad cars, for he has been since connected for years with the running of them. The observation of them has become his pursuit in life. It was natural, therefore, for him to observe and recollect such a fact; and a boy of that kind, with such a predisposition, could remember that he saw some of the rioters brought into Baltimore, on an eight-wheel car, with boards put on it; hence there is strong probability that McClarie is correct, particularly as the men who manufactured the cars testify to their having been in existence before and at that time.

By the testimony of William E. Rutter, defendants' No. 3, pages 156 to 158—folios 624 to 632.

Mr. Rutter says that he saw these cars, in the winter of 1830, take cord-wood down the slope east from Mount Clare.

The plaintiff attempted to contradict that statement, by saying that there was no track laid there; but we prove that there was a track laid there in April and May, 1830, by the report of James P. Stabler, assistant engineer on the Baltimore and Ohio railroad, dated Oct. 1st, 1830, and contained in the 1st volume of reports of the company.

Mr. Rutter worked for Buddy and Colvin; and he says that they built some cars for the Baltimore and Ohio Railroad Company, before they had a shop of their own sufficiently large to do their business; and Wm. Woodville, auditor and master of transportation of the Baltimore and Ohio road, and one of the plaintiff's witnesses, fully supports him in this, at plaintiffs' proofs, page 32—folios 137, 138.

Mr. Rutter also says *that he is positive that he saw these eight-wheel wood cars running on the Baltimore and Ohio road in* 1830, *before the Columbus was commenced;* and he recollects when she was begun, was in the habit of going where she was being constructed, and knew the men who built her.

This is positive and direct evidence; and how can it be possible—as the plaintiff's witnesses assert—that these wood cars were not there until 1835?

Conduce Gatch, who left the road in May, 1834, testifies to the same fact; so that it is utterly incredible and impossible that the plaintiff's witnesses are correct when they say, and simply from recollection, that they were not there until 1835. The very men who made them left the employ of the company previous to that time, and they cannot be mistaken.

Mr. Rutter was entirely independent of the company; but his mind had a strong inclination to observe railroad machinery, as his after life shows, and he testifies positively that he saw them before the Columbus was commenced; that they were the first eight-wheel cars that he had seen in his life, and that he left Baltimore in the summer of 1832.

By the testimony of Edward Gillingham, Defendants' No. 4, pages 56, 57, 58—folios 214 to 222.

This gentleman is the son of George Gillingham, who was the superintendent of machinery of the Baltimore and Ohio road, and since dead; and, although we have lost what he knew upon the subject, still we have the positive testimony of his son that these eight-wheel wood cars were on the road in 1830, and before the Columbus. He describes them of the same construction as do the others of our witnesses; i. e., two four-wheel trucks, having king-bolts and bolsters, with string-pieces bolted to the bolsters, and standards at the ends of the string-pieces. He saw these wood cars there in 1830, and recollects them distinctly. His father had a shop there, and did work for the company; and he was often there, assisting in mechanical duties. His mind was on mechanical subjects, and he cannot, therefore, be mistaken.

He is now in the employ of the Baltimore and Ohio Railroad Company, as a clerk in the machinery department; is a gentleman of much respectability; one whose observations on such a subject enable him to determine how these wood cars were constructed, and the date at which they existed; which he testifies was in 1830, and before the Columbus was built, or even thought of.

By the testimony of John Rupp, Defendants' No. 3, pages 43, 44—folios 172 to 179.

John Rupp resides in Hanover, Pennsylvania. He went to Baltimore in April, 1831, worked on the Columbus, and when he arrived there these eight-wheel wood cars were in existence on the Baltimore and Ohio road, and he, of course, as he testifies, saw them at that time. He left in 1832, and returned to Hanover, and cannot, therefore, be mistaken.

The entries in the time book of Gatch prove Mr. Rupp's statements as to time, showing that he came in April, 1831; which shows that the plaintiff's witnesses cannot be correct in drawing the inference that they were not there until 1835, simply because they do not recollect them, or did not particularly observe them.

By the testimony of John M. Eichelberger, Defendants' No. 4, page 21—folio 80; and page 37—folio 139.

Mr. Eichelberger was engaged as a painter in the service of the Baltimore and Ohio Railroad Company, and recollects these wood cars. There was no question in his mind but that they were eight-wheel wood cars, and that they hauled cord-wood.

I next refer to the extract from the Baltimore Gazette, of date Wednesday, January 19, 1831. " The Railroad is now furnishing," &c., &c., to be found in defendants' No. 4, pages 45. 46—folios 172 to 174, inclusive. That newspaper makes a matter of record on January 19th, 1831, of the fact " that a conveyance for wood had been furnished by the Baltimore and Ohio Railroad Company;" that the effect had been to keep the price of wood down that winter; that they would continue to do so, from the quantity furnished; and that the cars were exactly suited to this purpose.

By the testimony of James B. Dorsey, defendants' No. 4, page 75—folios 282 to 285, and 289; and page 77—folios 277, 278, 291 to 296.

Mr. Dorsey testifies that he went up to Frederick, in the latter part of 1831, and remembers examining one of these wood cars, which had been " constructed in Frederick, under the direction of Joshua Dill, the agent of the railroad company;" so that the building of these cars was not confined to Mount Clare, but they were so well known that one, with a railing around, was put together some sixty miles from Baltimore, in 1832, and was seen by Mr. Dorsey, and when he came to Baltimore he saw on the road large numbers of these eight-wheel wood cars, fixed with string-pieces; they were then a common car, in extensive use. Neither can he be mistaken in the time; for he left in 1832, and went to Columbia, and thence on the Frenchtown road. The plaintiff's witnesses, therefore, must be in error.

I next call attention to the second volume and sixth annual report of the president and directors to the stockholders of the Baltimore and Ohio Railroad Company; and therein, to the report of William Woodville, superintendent of transportation, dated 1st October, 1832, commencing: "Sir, the period," &c., &c.: " By reference to these papers," &c., &c. And, in the table "*M*, No. 7," to the article " 4480 *tons of firewood*," &c., under the entry of transportation, as wood carried during the winter of 1831.

I have shown, by a public newspaper as well as by witnesses, that wood was transported in 1830. And although there was no account of it kept and reported in 1830, and from the 1st of January, 1831, still, in corroboration, we find in the reports that, from the 1st October, 1831, to the 1st October, 1832, the large amount of " 4480 *tons* of firewood " were transported.

The testimony of Gatch, Rupp and Gillingham shows that more of these wood cars were constructed in 1831; that they then carried double tiers of wood; that it was a very cold and inclement winter ; and, that four-wheeled cars were also pressed into use, in consequence thereof, some of them having standards set up in the boxes, and others, pieces laid on the platforms, and fixed or fastened in that way.

All these facts, so far as they go, confirm, fully and entirely, the evidence of the defendants' witnesses—that eight-wheel cars commenced carrying wood in the fall of 1830.

By the testimony of Edward May, defendants' No. 3, page 46—folios 181, 184, 186 to 188.

Mr. May is now a merchant, and a hardware dealer in the city of Baltimore. He was, originally, a carpenter by trade and occupation; has been very successful, and is an intelligent and respectable man. He went to the Charles-street shop in the latter part of 1831, to assist in fitting up the shop ; and did assist in putting new standards in these wood cars. The track was laid from the Charles-street shop, running up Pratt-street, to Mount Clare. It had been laid in Pratt-street before; it ran in the direction of Mt. Clare. This track was laid from the corner of Charles and Camden, up to Pratt streets, in the early part of 1832; and Mr. May there saw some 15 or 20 of these wood cars, during the exceedingly cold winter, in which the greatest demand for them occurred. They had been brought there for the purpose of having new standards put to them, which repairs Mr. May did himself. He left in 1833; and one of the eight-wheel cars—the Winchester or the Dromedary—was then being constructed. In the spring of 1832, he testifies to these wood cars being there. He assisted in putting standards on them, and they were old cars; they had the appearance of having been in use a long time. It was his business, not only to see and observe, but to understand their construction, organization and their condition. It was perfectly natural for him to do so, because it related in a measure to the history of his life as connected with that road; it was something he had done and observed, and bore a close relation to his regular business. Men are more likely to remember that which is intimately connected with their business, and their own acts, than those who, from mere observation, testify as to what their recollection was or was not, in regard to events in which they were not actually concerned.

By the testimony of Henry Shultz, defendants' No. 4, pages 40 and 42—folios 154, 157 to 159 inclusive, who was there with Mr. May, and states the same facts as to the cars.

I next refer to the report of James P. Stabler, assistant engineer to Jonathan Knight, chief engineer, dated September 30th, 1830. On page 81 of the 4th annual report of the president and directors to the stockholders of the Baltimore and Ohio Railroad Company, 1st volume of reports, commencing with the words—"After the commencement," &c., Mr. Stabler there reports that a large quantity of these string-pieces, which the witnesses testify they used to construct these wood cars, was deposited near Mount Clare and

Pratt-street; and that it was then thought they might not be required for building the road, as stone sills were to be used. And hence the string-pieces, being so well suited for the construction of these cars, were at hand for consumption, and a few of them were used for that purpose.

I also refer to the 6th annual report, volume 2, pages 75 to 82. On page 75, to "office of construction," &c., "Sept. 30, 1832," from "the following report," &c., as far as "from the commencement of the work." On page 78, to "scantling procured for the railway has been used for building, for bridges, *for various parts of the machinery connected with the road*, and for a variety of other purposes." On page 79, to "the whole amount," &c., and from "6 × 6 inch yellow pine and other scantling," &c. On page 82, to "leaving 64,260 *feet appropriated to other purposes, as before mentioned.*"

That report of Mr. Stabler shows that 64,260 feet of this timber was used for *machinery* and bridges, &c., and hence, the plaintiff's assertion, that the entries on the pay rolls should have shown the purchase and use of these string-pieces, is unfounded. The string-pieces were there, already purchased by the superintendent of construction, and the work was rough, out of door work; they were put together out on the track, and hence Mr. Gatch very properly did not report them as new cars, nor as new stock. There was no report of such work needed, and none was made; and he could not have been expected to have reported them as lumber, purchased by him for the company, when he did not purchase it; it was already in their possession. All these witnesses whom I have referred to, prove clearly and fully that the mere change, occupying but a few hours, of bolting these string-pieces fast to the bolsters of the timber cars, and allowing them to remain permanently in that way, was done in 1830, and that cordwood was transported upon these wood cars prior to the commencement of the Columbus; so that the plaintiff's witnesses must be mistaken in the inference which the plaintiff attempts to draw from their testimony, because they do not recollect that the cars were there until 1835, whilst the very men who built them had left the service of the company, many of them long before that time. They were certainly there and in use in November, 1830.

Trussel Cars, 1830.

My next object is to prove that the eight-wheel, double truck, trussel cars were made, and in use for transporting horses and a carriage together in the same trussel, and upon eight wheels, at least as early as December 17th, 1830, being prior to the planning and building of any eight-wheel passenger car; and for which I rely upon the following witnesses:

Jacob Rupp, defendants' No. 4, pages 63 to 68—folios 141 to 264; Jacob Rupp, defendants' No. 4, pages 61, 62—folios 237 to 239; Leonard Forrest, defendants' No. 4, page 38—folios 144 to 146; Jno. M. Eichelberger, defendants' No. 4, page 37—folio 139; Edward Gillingham, defendants' No. 4, page 56—folios 214 to 219; Conduce Gatch, defendants' No. 4, page 35—folios 132 to 134; Conduce Gatch, plaintiff's proofs, page 104—folios 495 to 497; "Exhibit B," plaintiff's proofs, page 109; containing "Two trussels for the transportation of horses and carriages;" Extract from "Baltimore American," date December 18, 1830; Defendants' No. 4, page 55—folios 210 to 213; Extract from "Baltimore Gazette," date December 17, 1830; and, Jno. S. Sumner, defendants' No. 4, page 48—folios 184 to 186; John Rupp, defendants' No. 3, page 43—folio 172; John Rupp, defendants' No. 3, page 44—folio 174.

Mr. Jno. Rupp assisted to fix up a car, similar in description, in 1831; but all these other witnesses, and the entries, show clearly that in December, 1830, two of these eight-wheel trussel cars, with a framing around, and end pieces to let down with hinges, were constructed and in use; that they transported the carriage and horses of John Quincy Adams; and that the construction of the trussel part was made a matter of record upon the pay-rolls, because it required considerable work, was an addition of new stock, and comparatively expensive.

The newspaper publications fixed the date fully, and the descriptions contained in them corroborate the witnesses who built the cars.

Our witnesses testify that it was one long body (the trussel) resting upon two ordinary four-wheel trucks, which were well suited for such a purpose; that they were strong and comparatively small in size; that the carriage and horses were put in this trussel or car, and thus taken on the road. The pay-rolls show that two were built, and the newspapers show that more than one was built; that one trussel carried the carriage and horses; that it was hitched behind one of the passenger cars, in which some of the directors and Mr. Adams were seated; and that it was a novel and convenient mode of transporting horses and carriages, as they could be driven up the slopes that let down on to the platform, or trussel, without being unloaded.

The plaintiff's witnesses say that they nailed some rough pieces of board on two four-wheeled cars, to carry the carriage and horses of Mr. Adams; if they did nail such boards on such cars, they are mistaken as to the person, and it must have been before that time.

It is not at all probable that a mere casual rough piece of work, as that would be, was reported in the newspapers; it is utterly inconsistent; it cannot be believed; for they were very different cars described in the papers and reported in the pay-rolls.

The meaning of the word "*trussel*," is a body supported on two objects; which shows it to have been an eight-wheel car—consisting of two four-wheel trucks—to the mechanical mind, and it was very natural to have organized such a construction, after the timber and wood cars.

There was, however, no new principle about it, and no invention to form the basis of a patent, especially not by the plaintiff, who was not the originator of it.

The plaintiff alleges that there could not have been any wood or trussel cars on the road before the appearance of the Columbus, in July, 1831, and relies upon the following statements:

1st. That the officers and others connected with the road, in more or less responsible positions, do not remember seeing such cars until 1835.

2d. That the said trussel and wood cars were not specifically mentioned in the reports as eight-wheel cars.

3d. That no specified charge appears in relation to them in the pay-rolls of the company.

In answer to the first proposition, I say that Philip E. Thomas does not dispute their existence, and he was president of the road; neither does George Brown, the treasurer; John Ferry, the clerk; George Maxwell, a machinist; Lloyd Claridge, a painter; and Thomas Davis, a conductor; and all these are the plaintiff's witnesses. Edward May, plaintiff's witness, page 124 plaintiff's proofs, folio 542, says that in "1832 *he saw the wood cars,*" and thus contradicts others of the plaintiff's witnesses, who say they were not there until 1835. William Woodville, another of the plaintiff's witnesses, does not dispute the existence of the trussel cars; and all the other witnesses of the plaintiff were not concerned in building such cars; Glenn was laying down switches, Cromwell was a coachmaker, and Woodville was keeping the transportation books; and besides which, it does not appear that he had a mechanical mind, and powers of observation and discrimination in mechanics, sufficient to identify a change as to principle in any cars whatever.

In reply to the second proposition, I say that the car Columbus was in existence in July, 1831, on eight wheels, and no mention is made of her in the reports between that time and 1834, although she cost a considerable sum of money; still she was there on the road, and so were the others; and, if she were not noticed, it is natural to suppose that the wood and trussel cars, which were but of trifling expense, comparatively, should be omitted also, although they were there.

In answer to the third proposition, I say that the trussel cars are reported in December, 1830, on the pay-roll; that the timber for the wood cars was purchased for the construction of the road, by the superintendent of construction, and is stated to be at the depot, September 30th, 1830, in the report of James P. Stabler.

James P. Stabler, in his 6th annual report, accounts for all this timber since the commencement of the construction of the road; and reports that 64,260 feet, running measure, were used for bridges *and machinery on the road.*

Judge NELSON.—These wood cars, of which you speak, are of no moment in the case, unless it can be shown that they were brought out before the Columbus; it is of no importance to show that they were there after the Columbus.

Mr. HUBBELL.—We show, conclusively, may it please the court, that they were there as early as November, 1830, which is long prior to the existence of the Columbus. Their use, after the Columbus, goes to the question of abandonment.

The pay-rolls that the plaintiff produces do not show work done on the Columbus; and neither was such work done on the wood cars in the shop. It was done out on the road, as the testimony of Conduce Gatch shows. It is also a reasonable supposition that, had such work been done in the shop, it would not have made its appearance on the pay-rolls, as it was not attended with an outlay of money by Gatch.

"Winans' cars," as reported, were "friction wheel cars." This gave rise to the name, "Winans' cars," being frequently used. The appellation, "Winans' cars," did not refer to the eight wheel cars; and whenever "Winans' cars" were mentioned by Gatch and others, it meant the friction wheel cars, and not the eight-wheel cars. See Gatch's pay-roll, page 111, plaintiff's proofs for November, 1831, before the Columbus was thought of.

Columbus.

I will now advert briefly to the Columbus; and, although Ross Winans probably produced a sketch or plan of the body framing of the car Columbus, without any bolsters or running gear, yet, that the person who made the arrangement of the running gear, which was merely repeating what had before been done upon the wood and trussel cars, was Conduce Gatch. To prove this, we rely upon Conduce Gatch, Troy case, pages 33, 34, from " In the first passenger car," &c., down to " date of Mr. Winans' patent." The time being corrected through his time book entries, defendants' No. 4, page 35 — folio 131; defendants' No. 1, page 29 — folios 185, 186; defendants' No. 1, page 30 — folios 191, 192, 193; defendants' No. 2, pages 21, 22 — folios 85, 86; defendants' No. 3, pages 25, 26 — folios 94 to 101; defendants' No. 3, page 27 — folios 104, 105.

John M. Eichelberger, defendants' No. 2, pages 23, 24 — folios 94 to 96; defendants' No. 4, pages 20, 21 — folios 75 to 80.

Jacob Rupp, defendants' No. 2, pages 15, 16 — folios 59 to 64; defendants' No. 4, pages 22, 23 — folios 84 to 89.

Leonard Forrest, defendants' No. 2, pages 18, 19 — folios 74 to 76, 77; defendants' No. 4, pages 17, 18, 19 — folios 64 to 72.

John Rupp, defendants' No. 3, pages 43, 44 — folios 172 to 179.

William E. Rutter, defendants' No. 3, pages 157, 158 — folios 626, 631, 632.

Edward Gillingham, defendants' No. 4, pages 57, 58 — folios 219, 220 to 222.

John P. Mittan, defendants' No. 4, page 58 — folios 223 to 225.

Henry Shultz, defendants' No. 4, page 41 — folio 157.

Thomas Murphy, defendants' No. 4, page 40 — folios 152, 153.

Extract from " Baltimore American," of date July 4, 1831.

John S. Sumner, defendants' No. 4, page 47 — folios 179, 180.

Extract from " Baltimore Gazette," of date July 2d, 1831.

Many of these witnesses swear positively that they saw the drawing from which the Columbus was built; and that it was simply the body framing, similar to the exhibit which was handed to the court with their testimony, without boards nailed upon it, and without finish, without bolsters and without trucks. They testify positively to that representing the drawing from which the Columbus was built, as to the body. They also testify that Gatch drew the running gear on a board, similar to the drawing handed to the court; and from it the trucks were built.

In the building of the Columbus, it is manifest, from the history of the development of this principle on the Baltimore and Ohio road, that it was unnecessary to give Gatch any directions as to the passenger cars; for all he wanted was the size and dimensions of the body, which was precisely what the drawing contained. He could ascertain, by laying his rule on it, whether the scale was $\frac{3}{4}$ or $\frac{1}{2}$ an inch to a foot. He could tell that it was not $\frac{1}{2}$ an inch, because that would be 36 feet long and 10 feet high, and entirely out of proportion.

In the Troy case, one of the witnesses says that he saw him measure it, and so he did; and the drawing being 18 inches long and 5 inches high, with a scale of $\frac{3}{4}$ of an inch to a foot, would make it 24 feet long and $6\frac{2}{3}$ feet high, which is about the proper size; and that was the size of the body. He measured it to obtain this scale and size. He had fixed the running gear before on similar cars. It was no trouble for him to make a similar drawing of the two trucks; which he did, and from which the men worked in building the car. There was no trouble about it, nor was it anything new. In that way, the Columbus was planned and built; and when she appeared on the road she was not spoken of by the papers of the day, nor by the reports of the company, as a new principle, but simply as a large car, on eight wheels, for passengers.

Winans' friction wheel is spoken of in the company's reports as a new and important invention; and if Winans was the inventor of the Columbus, why was she not spoken of as his, and something new and important as an invention, and not simply as a large and capacious car on eight wheels? Winans himself did not consider that he had invented her, and never has sworn that he did. All that we find recorded in relation to her is consistent with what our testimony shows was done before, because she did not embody any novel principle or material change of organization.

As the court has heard the testimony on this point, I will not consume time by repeating it, but will pass to another part of the same subject.

The plaintiff, in order to prove that Winans really planned and contrived the arrangement of the running gear of the car Columbus, relies upon two species of testimony:

First.—Witnesses.

Second.—*A smoked drawing.*

In answer to the plaintiff's witnesses, the defendants rely upon the following testimony; and in answer to the plaintiff's alleged drawing, the defendants prove that the car Columbus was not built in conformity with the alleged drawing, in any material particular, but was produced according to defendants' model presented to the court.

The defendants' witnesses are James B. Dorsey, defendants' No. 4, pages 76, 77—folios 287 to 291; John M. Eichelberger, defendants' No. 4, pages 36, 37—folios 135 to 139; Leonard Forrest, defendants' No. 4, pages 37, 38—folios 140 to 144; John P. McJilton, defendants' No. 4, page 39—folios 147 to 150; Extract from " Baltimore Patriot," of date June 24th, 1831: " Surrounded by an IRON railing," &c.; Henry Shultz, defendants' No. 4, page 41—folios 155 to 158; John P. Mittan, defendants' No. 4, pages 58, 59—folios 228 to 230; Jacob Rupp, defendants' No. 4, pages 60, 61—folios 231 to 239; Edward Gillingham, defendants' No. 4, pages 56, 57—folios 214 to 222; Conduce Gatch, defendants' No. 4, pages 29 to 34—folios 107 to 129.

The plaintiff's witnesses evidently speak, in their testimony, from their *general* recollection of the appearance of the Columbus, she being a yellow car, with trucks and perches, and the drawing is also painted yellow, displaying trucks and perches. They do not specify any details either in physics or mechanics sufficient to prove that they recollect the construction, or the manner in which the pieces were connected with each other. And when they testify to the best of their remembrance, they are very likely to be mistaken in regard to the identity of the drawing with the Columbus. That species of evidence, on a matter of this kind, cannot prevail against a close examination into the mechanics. The question whether a certain car was built from a certain drawing, cannot be determined from a *general* glance at a drawing, or recollection of the car. It can only be determined by a close comparison of the mechanical structure and arrangement, the mode of fixing the parts together, and their formation. And it is upon such an examination that the defendants' witnesses show that the drawing now before the court, and produced by the plaintiff, was never seen by them, and was not the design from which the Columbus was constructed.

The defendants' witnesses gave particulars; they say that the trucks in the drawing are not like it; that they are friction wheel boxes, and that the Columbus had not, *when first built*, friction wheel boxes, *but she had plain boxes;* that she *subsequently* had friction wheel boxes, *which, not answering, plain boxes were again put on the trucks.* Leonard Forrest made the bolts for the plain boxes when she was first built, and Conduce Gatch testifies that she had plain boxes on at first; that she was afterwards altered, and friction wheel boxes put on; that the trucks had to be constructed to suit the alteration; that they were tried, and after finding they would not answer, they were removed, and plain boxes were again put on. The witnesses do not contradict each other, but those on behalf of the plaintiff fall short of the statement of all the facts.

Mr. Gillingham, who is the son of the person who made these friction wheel boxes, testifies that she had plain boxes on when first built; that subsequently the friction wheel boxes were made at his father's establishment, and put on in 1832, at the Charles-street shop.

The smoked drawing has not the same construction of trucks that the Columbus at first had; the timbers are not put together in the same way; the perch is not of the same formation as when she was first built; she then had a *straight perch, with hownes.* This perch on the drawing is *tapered*, and has *no hownes.* The drawing shows friction wheel boxes; whilst she had plain boxes when first built. The drawing shows six sliding windows on a side, three feet by two feet each; she had nine or ten sliding windows on a side, each about two feet square. Such windows as the drawing shows could not have been put there with such a frame—the number of uprights is too few; the car was built with more uprights on a side. The bottom framing is not the same; they are not placed in the same way. It has only two longitudinal timbers; the Columbus had *four;* she had middle timbers running lengthways to support the floor. The side pieces did not terminate flush with the end pieces, as appears in this smoked drawing — they projected beyond the end pieces. This smoked drawing also shows the uprights continuing up through the roof, thus making a wooden railing; whilst there never was a wooden railing on the Columbus, and the dimensions of the panes of glass over each shutter were not so large; also the shutters and panes of glass were more numerous. It is utter folly to say that the Columbus was built from that drawing, as it is not, in the first place, even a working drawing. A drawing from which a car is to be built must be a drawing in detail — such a one as Mr. Gatch made, and such a one as was handed to him; and when the plaintiff attempts to support this drawing by showing that the Columbus was built from it, it is utterly an impossibility. In the drawing, there is no eave to the roof; the Columbus had an eave to turn off the rain; it is said also that the Columbus carried great weight on the top; and is it to be supposed that in such case the side pieces would be run up through the roof, as shown in this drawing, instead of the ends supporting it, as exhibited by the model? *They are alike only in the yellow and drab paint on the body and doors.* There are no thorough brace bolsters in the drawing, and the Columbus had them. It does not approach her construction in 1831; and when a comparison is made between them, the attempt utterly fails, and seventeen material mechanical differences, which tell no false tales, are found to exist, every one of them sufficient to prove that she was not built from the smoked drawing.

They are enumerated and fully stated in the following testimony: page 129—folio 109—defendants' No. 4:

1st. This drawing has the bottom side pieces of the body frame flush with the box part of the body at the ends.

The car Columbus had these side pieces to project about four inches beyond the box part of the body at each end, and rounded off like the model.

2d. This drawing has a body representing two doors on the same side, one at each end of the side. The car Columbus had one door at one end and a pair of steps at the other end of the same side of the body, and the two doors, and the two pair of steps were placed, the doors diagonally to each other, and the steps diagonally to each other, as shown in the model.

3d. This drawing has seven side, and two end uprights, showing two doors and six sliding windows on a side. The dimensions of the open space of the sliding windows is one and a half inch high, by two and a quarter inches long, which, at three-quarters of an inch to the foot, would be two feet high and three feet long.

The car Columbus had ten or eleven side uprights and two end uprights, forming nine or ten sliding windows, one door and a panel for a pair of steps on a side; and the dimensions of the open spaces of the sliding windows was about two feet square, like the model.

4th. This drawing has panes of glass above the sliding windows, which measure five-eights of an inch broad by three-quarters of an inch high, which, at three-quarters of an inch to the foot, would be ten by twelve inch panes of glass.

The car Columbus had panes of glass above the sliding windows about seven by nine inches, certainly not larger than seven by nines, and instead of being only eighteen in number over the slides, there were in the car twenty-seven or thirty in number on each side.

5th. This drawing represents wooden uprights of the same size as the uprights of the body, and a wooden railing with net work on top of the body.

The car Columbus never had a wooden railing on top. When she first came out on the road she had iron uprights, with a scroll and iron baggage rods on top, as shown by the model; and in the spring of 1832, she had iron uprights put on her to secure an awning to, and a wire net-work secured to these iron uprights, for the convenience of passengers.

6th. This drawing represents the top of the car as without any eave to project over the top side framing pieces.

The car Columbus had an eave projecting over each side of the body, to throw the rain off clear of the sides of the body.

In my opinion, no mechanic, who understood his business, would build a car with a roof to be supported inside of the framing, as indicated by the drawing, as it would neither throw the water off at the sides nor rest on a firm basis; it would not properly sustain a heavy load of baggage or passengers. The roof of the car Columbus rested on top of the framing of the body.

7th. This drawing has no opening at each end through the railing there shown, for passengers or persons to pass through, and no visible means for persons or baggage to be taken on top. The car Columbus had two iron uprights, tipped with brass balls at the head of the steps at each end, and ropes from them for persons to ascend the steps and pass through on top, and these formed a finish to the baggage rods, all as shown by the model.

8th. This drawing measures four and a quarter inches from the middle of the bottom framing pieces to the top framing pieces, which, at three-quarters of an inch to the foot, would give five feet eight inches as the height from the floor to the top, inside. This would be too low for an ordinary sized man to stand up in with his hat on.

The car Columbus was high enough inside for a tall man to stand up in with his hat on, and have it just touch the top, and was about six feet six inches high inside from the floor to the top.

9th. This drawing has the side framing piece immediately below the panes of glass, to measure on the scale of three-quarters of an inch to the foot, about three and half inches in width, and the top side framing piece to measure two inches in width. Two inches in width would not be sufficiently strong to support a roof safely, particularly when it is to carry a load on top, and the uprights are three feet apart, as in the drawing or lithograph.

The car Columbus had the uprights of the frame about two feet apart, and the side framing pieces at the top rested on the uprights; they were mortised into it, and the roof rested, with an eave projecting over the sides, on top of these upper side framing pieces; and these upper side framing pieces were wider and stronger than the framing pieces that were immediately below the glass, because the upper framing pieces had the weight of the roof and any load that might be placed on it to sustain, and therefore needed to be the largest and strongest.

10th. Referring now to the ground plan of the framing for a car body, shown in this drawing or lithograph, the same difference, as mentioned first, is seen in the side framing pieces not projecting beyond the end pieces. Also this ground plan has no middle pieces extending longitudinally and parallel with the side framing pieces to support the floor in the middle.

The car Columbus had two middle pieces, or timbers between the side pieces, extending longitudinally and parallel with the side pieces to support the floor of the body, as shown in the model.

11th. As to the running gear. This drawing shows bolsters at a distance from the centre of the bolsters to the end of the body, a little over one-fourth the length of the body.

The car Columbus, as built by me, had the bolsters distant from the ends of the body to their centres each one-fifth the length of the body. I recollect distinctly that I placed them at one-fifth, so that two-fifths should balance each other; one-fifth on each side of each bolster, and the remaining fifth in the middle of the body should be sustained by the main strength of the body itself.

12th. This drawing or lithograph shows a plain bolster to the trucks and body, and does not show any king-bolt.

The car Columbus had a thorough brace bolster and king-bolt to each truck, as shown in the model.

13th. This drawing does not show any side bearing friction rollers to enable the trucks to swivel easily.

The car Columbus had side bearing friction rollers in the truck bolsters, two to each truck, as shown in the model.

14th. This drawing shows perches, tapered and passing directly through the body of the truck bolsters, and without any hownes or braces.

The perches of the car Columbus were not tapered; they were straight, and did not pass directly through the body of the truck bolsters; they were pillowed upon and notched into the top of the truck bolsters, as shown in the model, and had hownes.

15th. The car Columbus had the end pieces and bolsters of the trucks notched into the tops of the side pieces, and an iron rod or bolt passing under each bolster, securing the side pieces; and the side pieces were rounded at the ends, horizontally, all as shown in the model.

This drawing has no iron rod or bolt under the truck bolsters to secure the side pieces together; has the bolsters and end pieces pillowed on the side pieces, but not notched into them; and has the ends of the side pieces rounded vertically, not horizontally.

16th. This drawing shows the large semi-circular boxes on the trucks that were used to inclose Winans' friction wheels, in the latter part of 1831, on some four-wheeled cars, and in the early part of 1832 were tried with Winans' friction wheels, chilled on the inner surface, on the Columbus, for which she had to have new truck-frames made, with the side pieces, consisting of two timbers, bolted together, and a curved mortise in them for the friction wheels to run in, and thus protect them from dust as much as possible; the side pieces of the first trucks put under her were single timbers, and not such as the inclosed Winans' friction wheels were afterwards applied to. The journals of the friction wheels did not, though, come through the metal of the box inclosing them, as appears in the drawing; that is not mechanical; the journals of the friction wheels rested in boxes bolted to the side pieces, inside of the boxes, such as shown in the drawing, and the sides of these large boxes were solid, and inclosed the under side of the friction wheel and its bearings. The chilling of the inner surface of the friction wheels and inclosing them in these boxes was, at that time, thought to be an improvement worth a trial; but owing to the accumulation of dust in the box, by passing through the hole where the axle passed into the friction wheel, and the cutting of the metal, it was found impossible to make the friction wheels answer, and they were abandoned generally, and plain boxes were again put on the Columbus. The car Columbus, when first built, had small plain boxes on her, and the side pieces of the trucks were single timbers, like the model.

17th. This drawing or lithograph, in the ground plan, shows a body cross piece or bolster to the framing, and then a shorter bolster, having a ring to pass between it and the body timber or bolster. This ring passing between them indicates that they move upon each other, or that the shorter one is the truck bolster. This position of the ring between them will not, in mechanics, admit of any other conclusion than that the shortest bolster is a truck bolster. This truck bolster measures, on the lithograph, three inches and nine-sixteenths, which, at three-quarters of an inch to the foot, would make it 57 inches long, being the same length as the guage of the track of the Baltimore and Ohio railroad, and wholly insufficient to reach across the truck from one side piece to the other of such trucks as the Columbus was built with, as the side pieces were outside of the wheels, like the model.

The car Columbus had truck bolsters about seven feet long, and also had a safety coupling rod between the trucks, like the model, which this drawing has not.

It is proved, seventeen times over, therefore, that the position of the plaintiff is incorrect, and that the drawing had no existence at that time, and of course was subsequently made, in 1832–'33 or '34. We prove, by two processes, that the drawing did not exist at the time, but that it was subsequently made, with friction boxes corresponding, in a measure, with those under the Columbus in 1832. The draughtsman who made that smoked drawing has not been particular enough to make the body correspond with the Columbus, or with the skeleton drawing which was handed to Gatch in 1831. The principle was not new, and Gatch needed nothing more than the outline of the body. The plaintiff's witnesses are entirely mistaken in their recollection or belief on the subject.

Winans himself was engaged in engineering pursuits and iron machinery, experimenting with chilled and friction wheels, planning machinery for the inclined planes, and looking after the steam engines, in 1831. And for reference, in relation to the manner in which he was engaged, I will, in addition, call the attention of the court to the fifth annual report of the company, in 1831. To the report of John Elgar, of date September 30, 1831, pages 99, 100. To the report of Jonathan Knight, of date October 1st, 1831, page 22, vol. 1, of reports. Also, to the report of Jonathan Knight, of 1833, pages 36, 37, 38. Also, the report of Jonathan Knight, of 1834, pages 26, 27; also to page 28.

In all these reports, there is not a single statement made from which the inference can be drawn that Winans was experimenting on eight-wheel cars. I also refer to the report of Geo. Gillingham, dated Oct. 1st, 1834, in which he mentioned the Winchester, the Dromedary, the Comet, and four others, made out of four-wheel cars; but he does not pretend to say that Winans was the inventor of them. The reason is evident, i. e., it was not for a moment openly pretended that he invented the eight-wheel car, and the engineers consequently did not so report; but had it been so, as they were in the habit of stating how Winans was employed, the fact would have appeared prominently in the reports before the 1st of

October, 1834 ; and as it is not mentioned, it not only shows that Winans was not the inventor, but it proves that it was in general use on the road, and that the principle, in this country and in England, was old.

Conduce Gatch left the employ of the company in May, 1834; *and it appears that Winans' long cherished hope and folly, the friction wheel, had utterly failed in* 1833 *and* 1834, *and as his fortunes were sinking, and Gatch was away, he conceived the idea of getting up a specification for an eight-wheel car.* He thought somebody might get some sort of a patent, and he did so, in the language in which it now appears, apparently avoiding the wood and trussel cars, and Columbus, previously constructed.

According to our understanding of it, it does not include or comprehend the Columbus, in her original construction ; it does not include the wood cars ; it does not include the trussel cars ; neither does it include the cars made by the defendants ; but if it be construed to include them, then we say the specification and patent are too broad.

We have exercised our best judgments in regard to the extent of Winans' patent; and we have arrived at the conclusion in our own minds, that its phraseology and conditions, and peculiar features, have required us to present the testimony in the manner in which it has been submitted. Of course, we have no right to pre-suppose that this court will decide in any particular way, and therefore we rely upon an alternative, viz : if Winans' patent be so broad as to cover the rigid wheel-frame eight-wheel cars, with the wheels say fifteen or twenty inches apart, and not " as close together as possible, to nearly coincide with the mean radial line," then it is too broad, and void in consequence thereof.

That is our position in relation to the wood cars as well as the others, and we contend that they in 1830 were substantially the same as those now manufactured by the defendants and in general use, and we have offered the testimony of Messrs. Stewart and McAlpine, and many others, as showing that fact. We submit that we have proved fully :

1st. That the wood cars, or frame on eight wheels, used in 1830, are substantially the same as the defendants', and before the time of the plaintiff's alleged invention.

2d. That the trussel car on eight wheels, used to transport carriages and horses in December, 1830, is substantially the same as the defendants', and before the time of the plaintiff's alleged invention.

3d. That the eight-wheel car Columbus, contrived and built by Conduce Gatch, and in operation July 4, 1831, is substantially the same as the defendants', and before the time of the plaintiff's alleged invention.

The Comet, 1834.

The next and last point to which I will very briefly advert is, that the plaintiff's alleged invention was first embodied in the Comet, which was the last car built in 1834, before the date of his patent, and is described by his specification. The car Comet had the big springs, connecting the axles with the wheels, as close as could be without touching; there being a space of half an inch between them, and coupled under one body. It was different in its machinery, in that particular, from the Columbus, the wood and trussel and the other cars. He therefore intended to describe it in his specification, and does describe it; and such is the theory that he has embodied in it. He does not state whether she had inside or outside bearing, although I presume that his intention was to include both. There is no mode of drawing the car described, and consequently the mode of drawing by the body cannot be claimed.

Winans' Specification.

And in order to prove that the whole theory and machinery of Winans' specification is a fallacy and pernicious, I will refer to the following list of witnesses, with reference to their testimony on this important part of the case:

William B. Aitken, defendants' No. 2, page 5 — folio 19.

Henry Waterman, defendants' No. 3, pages 64 to 66 — folios 256 to 262.

Jacob Schryack, defendants' No. 3, page 40 — folios 157 to 159.

Wm. T. Ragland, defendants' No. 3, page 68 — folios 269 to 272.

Albert S. Adams, defendants' No. 3, pages 114, 115 — folios 456 to 458.

George Stark, defendants' No. 3, page 70 — folios 278 to 280.

Charles B. Stuart, defendants' No. 3, pages 89, 90 — folios 356, 357.

Walter McQueen, defendants'. No. 3, page 103 — folios 409 to 411.

George W. Smith, defendants' No. 3, pages 81, 82 — folios 321 to 327.

George S. Griggs, defendants' No. 3, pages 109, 110—folios 425 to 439.

Wm. J. McAlpine, defendants' No. 3, page 93 — folios 370, 371.

Wm. Raymond Lee, defendants' No. 3, page 111 — folios 441 to 443. In which he corrects his affidavit in defendants' No. 1.

James H. Anderson, defendants' No. 3, page 118 — folios 470, 471.

Asahel Durgan, defendants' No. 3, pages 114, 115 — folios 456 to 458.

John B. Winslow, defendants' No. 3, page 121 — folios 482, 483.

Henry W. Farley, defendants' No. 3, page 129 — folios 513 to 515.

John Crombie, defendants' No. 3, page 126 — folios 502 to 504.

Albert Bridges, defendants' No. 3, pages 134 to 136 — folios 536 to 542. In which he corrects his deposition in defendant's No. 1.

Robert Higham, defendants' No. 3, pages 133, 134 — folios 529 to 534. Also a correction of testimony in defendants' No. 1.

Isaac Adams, defendants' No. 3, pages 138, 139 — folios 551 to 560. Also a correction of affidavit in defendants' No. 1.

J. Van Rensselaer, defendants' No. 3, page 137 — folios 544 to 548. Also a correction of affidavit in defendants' No. 1.

Wm. C. Young, defendants' No. 3, page 150 — folios 598 to 600.

Wm. P. Parrott, defendants' No. 3, pages 153, 154 — folios 611 to 614.

Charles Minot, defendants' No. 3, page 159 — folio 637.

Godfrey B. King, defendants' No. 4, pages 81, 82 — folios 307 to 309.

Stephen Ustick, defendants' No. 2, page 6 — folios 21, 22.

Oliver Byrne, defendants' No. 2, pages 12, 13 — folios 47 to 50.

Edward Martin, defendants' No. 2, pages, 34, 35 — folios 143 to 145.

George Beach, defendants' No. 2, pages 36, 37 — folios 151 to 155.

Stephen W. Worden, defendants' No. 2, pages 39, 40 — folios 164 to 168.

Henry Shultz, defendants' No. 3, page 44 — folios 168 to 170.

These prove the doctrines and machinery of the specification false and pernicious.

I must thank your honor for the kind attention manifested, confident that I have to some extent wearied you—but with a consciousness that I have endeavored to progress in this matter as fully and carefully as possible—although, perhaps, I have fallen short of the extent that its importance deserves.

I will now leave it in the hands of my able colleague.

NOTE.

The following is the specification of Ross Winans, as embraced in his patent, upon which this suit is based:

THE UNITED STATES PATENT OFFICE.

To all persons to whom these presents shall come, Greeting:

THIS IS TO CERTIFY, That the annexed is a true copy from the Records of this office of Letters Patent, issued to Ross Winans on the first day of October, eighteen hundred and thirty-four.

IN TESTIMONY WHEREOF, I, Charles Mason, Commissioner of Patents, have caused the Seal of the Patent Office to be hereunto affixed this seventh day of [L. S.] June, in the year of our Lord one thousand eight hundred and fifty-three, and of the Independence of the United States the seventy-seventh.

CHARLES MASON.

THE UNITED STATES OF AMERICA.

To all to whom these Letters Patent shall come:

WHEREAS, Ross Winans, a citizen of the United States, hath alleged that he has invented a new and useful improvement in the construction of Cars or Carriages intended to run on Railroads, which improvement he states has not been known or used before his application; hath sworn that he does verily believe that he is the true inventor or discoverer of the said improvement; hath paid into the Treasury of the United States the sum of thirty dollars, delivered a receipt for the same, and presented a petition to the Secretary of State, signifying a desire of obtaining an exclusive property in the said improvement, and praying that a Patent may be granted for that purpose: *These are therefore* to grant, according to law, to the said Ross Winans, his heirs, administrators or assigns, for the term of fourteen years, from the first day of October, one thousand eight hundred and thirty-four, the full and exclusive right and liberty of making, constructing, using, and vending to others to be used, the said improvement: a description whereof is given in the words of the said Ross Winans himself, in the schedule hereto annexed, and is made a part of these presents.

IN TESTIMONY WHEREOF, I have caused these Letters to be made Patent, and the Seal of the United States to be hereunto affixed.

[L. S.] GIVEN under my hand, at the CITY OF WASHINGTON, this first day of October, in the year of our Lord one thousand eight hundred and thirty-four, and of the Independence of the United States of America the fifty-ninth.

By THE PRESIDENT: ANDREW JACKSON.
JOHN FORSYTH, *Secretary of State.*

CITY OF WASHINGTON, *to wit :*

I DO HEREBY CERTIFY, *That the foregoing Letters Patent were delivered to me on the first day of October, in the year of our Lord one thousand eight hundred and thirty-four, to be examined ; that I have examined the same, and find them conformable to law ; and I do hereby return the same to the* SECRETARY OF STATE *within fifteen days from the date aforesaid, to wit, on this first day of October, in the year aforesaid.*

B. F. BUTLER,
Attorney General of the United States.

The Schedule referred to in these Letters Patent, and making part of the same, containing a description in the words of the said Ross Winans himself of his improvement in the construction of cars or carriages intended to run upon railroads.

To ALL WHOM IT MAY CONCERN: Be it known that I, Ross Winans, civil engineer of the city of Baltimore, in the State of Maryland, have invented a new and useful improvement in the construction of cars or carriages intended to travel upon railroads, which improvement is particularly adapted to passenger cars, as will more fully appear by an exposition of the difficulties heretofore experienced in the running of such cars at high velocities, which exposition I think it best to give in this specification for the purpose of exemplifying the more clearly the object of my said improvement.

In the construction of all railroads in this country which extend to any considerable distance, it has been found necessary to admit of lateral curvatures, the radius of which is sometimes but a few hundred feet, and it becomes important therefore so to construct the cars as to enable them to overcome the difficulties presented by such curvatures, and to adapt them for running with the least friction practicable upon all parts of the road. The friction to which I now allude is that which arises from the contact between the flanches of the wheels and the rails, which, when it occurs, causes a great loss of power, and a rapid destruction of or injury to both the wheel and the rail, and is otherwise injurious.

The high velocities attained by the improvements made in locomotive engines, and which are not only sanctioned but demanded by public opinion, render it necessary that certain points of construction and arrangement, both in the roads and wheels, which were not viewed as important at former rates of traveling, should now receive special attention. The greater momentum of the load, and the intensity of the shocks and concussions, which are unavoidable even under the best constructions, are among those circumstances which must not be neglected; as the liability to accident is thereby not only greatly increased, but the consequences to be apprehended much more serious. The passenger and other cars in general use upon railroads have four wheels, the axles of which are placed from three and a half to five feet apart; this distance being governed by the nature of the road upon which they run, and other considerations. When the cars are so constructed that the axles retain their parallelism, and are at a considerable distance apart, there is a necessary tendency in the flanches of the wheel to come into contact with the rails, especially on the curvatures of least radius, as the axles then vary more from the direction of the radii. From this consideration, when taken alone, it would appear to be best to place the axles as near to each other as possible; thus causing them to approach more nearly to the direction of the radii of the curves and the planes of the wheels, to conform to the line of the rails. There are, however, other circumstances which must not be overlooked in their construction. I have already alluded to the increased force of the shocks from obstructions at high velocities; and whatever care may be taken, there will be inequalities in the rails and wheels, which, though small, are numerous, and the perpetual operation of which produces effects which cannot be disregarded. The greater the distance between the axles, while the length of the body remains the same, the less is the influence of these shocks or concussions; and this has led, in many instances, to the placing them in passenger cars at or near their extreme ends. Now, however, a compromise is most commonly made between the evils resulting from a considerable separation and a near approach, as by the modes of construction now in use, one of the advantages must be sacrificed to the other.

But it is not to the lateral curvatures and inequalities of the road alone that the foregoing remarks apply. The incessant vibration felt in traveling over a railroad is mainly dependent upon the vertical motion of the cars in surmounting those numerous, though minute obstructions which unavoidably exist. The nearer the axles are placed to each other, the greater is the effect of this motion upon the passengers, and the greater its power to derange the machinery and the road. It becomes very important, therefore, both as regards comfort, safety and economy, to devise a mode of combining the advantages derived from placing the axles at a considerable distance apart with those of allowing them to be situated near to each other. It has been attempted, and with some success, to correct the tendency of the flanches to come into contact with the rails on curved and other parts of the road, making the tread of the wheel conical. And if the traveling upon railroads was not required to be very rapid, this would so far prove an effectual corrective, as the two rails would find diameters upon the wheels which would correspond with the difference in length, the constant tendency to deviation being as constantly counterbalanced by this construction; but at high velocities, the momentum of the body in motion tends so powerfully to carry it in a right line as to cause the wheel on the longer rail to ascend considerably above that part of the cone which corresponds therewith. The consequence of this is a continued serpentine motion, principally, but not entirely, in a lateral direction. Nor is this confined to the curved parts of the road, but it exists to an equal or greater extent upon those which are straight, especially when the axles are near to each other; the irregularities before spoken of, constantly changing the direct course of the wheels, whilst there is no general curvature of the rails to counteract it. To avoid this effect, and the unpleasant motion and tendency to derangement consequent upon it, an additional motive is furnished for placing the axles at a considerable distance apart.

The object of my invention is, among other things, to make such an adjustment of the wheels and axles as shall cause the body of the car or carriage to pursue a more smooth, even, direct and safe course than it does, as cars are ordinarily constructed, both over the curved and straight parts of the road, by the beforementioned desideratum of combining the advantages of the near and distant coupling of the axles, and other means to be hereinafter described. For this purpose, I construct two bearing carriages, each with four wheels, which are to sustain the body of the passenger or other car, by placing one of them at or near each end of it, in a way to be presently described. The two wheels on either side of these carriages are to be placed very near to each other; the spaces between their flanches need be no greater than is necessary to prevent their contact with each other. These wheels I connect together by means of a very strong spring—say double the usual strength employed for ordinary cars—the ends of which spring are bolted, or otherwise secured, to the upper sides of the boxes, which rest on the journals of the axles; the longer leaves of the springs being placed downwards, and surmounted by the shorter leaves. Having thus connected two pairs of wheels together, I unite them into a four-wheel bearing carriage by means of their axles, and a bolster of the proper length, extending across, between two pairs of wheels, from the centre of one spring to that of the other, and securely fastened to the tops of them. This bolster must be of sufficient strength to bear a load upon its centre of four or five tons. Upon this first bolster I place another of equal strength, and connect the two together by a centre pin, or bolt, passing down through them, and thus allowing them to swivel or turn upon each other, in the manner of the front bolster of a common road wagon. I prefer making these bolsters of wrought or cast iron; wood, however, may be used. I prepare each of the bearing carriages in precisely the same way. The body of the passenger or other car I make of double the ordinary length of those which run on four wheels, and capable of carrying double their load. This body I place so as to rest its whole weight upon the two upper bolsters of the two beforementioned bearing carriages or running gear. I sometimes place these bolsters so far within the ends of the body of the car as to bring all the wheels under it, and in this case, less strength is necessary in the car body than when the bolster is situated at its extreme ends. In some cases, however, I place the bolster so far without the body of the car, at either end, as to allow the latter to hang down between the two sets of wheels, or bearing carriages, and to run, if desired, within a foot of the rails.

When this is done, a strong frame-work projects out from either end of the car or carriage body, and rests upon the upper bolsters of the two bearing carriages. This last arrangement, by which the body of the car is hung so low down, manifestly affords a great security to the passengers, exempting them, in a great degree, from those accidents to which they are liable when the load is raised. Several bodies may be connected, or rest on a common frame, and be supported on the bearing carriage, in a manner similar to that of a single body. When the bolsters of the bearing carriages are placed under the extreme ends of the body, the relief from shocks and concussions, and from lateral vibrations, is greater than it is when the bolsters are placed between the middle and the ends of the body, and this relief is not materially varied by increasing or diminishing the length of the body, while the extreme ends of it continue to rest on the bolster of the bearing cars, the load being supposed to be equally distributed over the entire length of the body.

Although I prefer the use of a single spring to a pair of wheels as above described, instead of the ordinary spring to each wheel, and consider it as more simple, cheap and convenient than any other arrangement; the end which I have in view may, nevertheless, be obtained by constructing the bearing carriages in any of the modes usually practiced, provided that the fore and hind wheels of each of them be placed very near together; because the closeness of the fore and hind wheels of each bearing carriage, taken in connection with the use of two bearing carriages coupled remotely from each other as can conveniently be done, for the support of one body, with a view to the objects and on the principles herein set forth, is considered by me as a most important feature of my invention, for by contiguity of the fore and hind wheels of each bearing carriage, while the two bearing carriages may be at any desirable distance apart, the lateral friction from the rubbing of the flanches against the rails is most effectually avoided, whilst, at the same time all the advantages attendant upon placing the axles of a four-wheeled car far apart, are thus obtained. The bearing of the load on the centre of the bolster, which also is the centre of each bearing carriage, likewise affords great relief from the shocks occasioned by the percussion of the wheels on protuberant parts of the rails or other objects, and from the vibrations consequent to the use of coned wheels; as the lateral and vertical movements of the body of the car resulting from the above causes, are much diminished. The two wheels on either side of one of the bearing carriages may, from their proximity, be considered as acting like a single wheel, and as these two bearing carriages may be placed at any distance from each other, consistent with the required strength of the body of the car, it is evident that all the advantage is obtained which results from having the two axles of a four-wheeled car at a distance from each other, whilst its inconveniences are avoided.

Another advantage of this car, compared with those in common use, and which is viewed by me as very important, is the increased safety offered by it to passengers, not only from the diminished liability to breakage, or derangement in the frame work, but also from the less disastrous consequences to be apprehended from the breaking of a wheel, axle, or other part of the running gear, as the car body depends for its support and safety upon a greater number of wheels and bearing points on the road. I do not claim as my invention the running of cars or carriages upon eight wheels, this having been previously done; not, however, in the manner or for the purposes herein described, but merely with a view of distributing the weight carried, more evenly upon a rail or other road, and for objects distinct in character from those which I have had in view, as hereinbefore set forth. Nor have the wheels, when thus increased in number, been so arranged and connected with each other, either by design or accident, as to accomplish this purpose. What I claim, therefore, as my invention, and for which I ask a patent, is, the before described manner of arranging and connecting the eight wheels, which constitute the two bearing carriages with a railroad car, so as to accomplish the end proposed by the means set forth, or by any others which are analogous and dependent upon the same principles.

ROSS WINANS.

G. BROWN,
JNO. H. B. LATROBE, } *Witnesses.*

ARGUMENT

OF

WILLIAM WHITING, Esq.,

IN THE CASE OF

ROSS WINANS v. ORSAMUS EATON et al.,

FOR AN ALLEGED

INFRINGEMENT OF HIS

PATENT FOR THE EIGHT-WHEEL RAILROAD CAR,

BEFORE

HON. SAMUEL NELSON,

JUSTICE OF THE UNITED STATES CIRCUIT COURT FOR THE

NORTHERN DISTRICT OF NEW YORK.

PHONOGRAPHICALLY REPORTED
By Arthur Cannon, of Philadelphia.

BOSTON:
J. M. HEWES & CO., PRINTERS, 81 CORNHILL.
1853.

ARGUMENT.

---•---

INJUNCTION.

MAY IT PLEASE YOUR HONOR:—

The great length to which this trial has been protracted, the unusual array of counsel, the thorough preparation of the evidence, and the elaborate arguments, which we have heard with so much pleasure, indicate that this cause is believed to be one of no ordinary importance. If success shall crown their enterprise, the parties really interested in the patent of Ross Winans, will be enabled to grasp a speculation worthy to be styled " magnificent."

Nor is the result of this struggle of less moment to the Respondents, Messrs. Eaton, Gilbert and Company. But however deeply the issue of this trial may affect the rights of the parties to the Record, their private contests sink into insignificance compared with the great public interests which must be affected by your decision.

Accidentally perusing a New York newspaper this morning, I observed the following paragraph:—

" The cost of one of the long railway passenger cars is, on an average, about $2,000. There are, in the United States, upwards of eighty private car manufactories, exclusive of those railways which make and repair all for their use, and it is calculated that a capital of $6,000,000 is invested in this branch of industry, producing about $17,000,000 annually, and employing about six thousand men."

Whether that is an accurate statement, or on what authority it is founded, we know not; but there is, beyond doubt, a vast amount of capital and many thousands of workmen employed in constructing eight-wheel cars for Railroad Companies in different sections of the country. The questions involved in these suits in equity, therefore, may be deemed, in no inconsiderable degree, as public questions. The same reasons which would authorize an injunction against the Respondents would require the Court to enjoin all the railroads of the State of New York; and will lay the foundation in every Circuit Court of the United States, for putting a stop to the railway travel in the cars now in general use. It is rarely that we have known, or been connected with any cause, which affects a greater variety of public and private interests than that which is now before this Court. And it is our good fortune to appear before a tribunal patient in investigating truth and anxious to administer justice, with even hand, no less to the unknown stranger than to the familiar friend. And if, indeed,—if a prayer had really been made for process to run throughout every State of this Union, and that prayer were now under consideration, this Court would feel no higher responsibility than it now feels in pronouncing judgment in this case, and thus giving a direction to the judicial mind of the country.

This cause is important in another aspect; namely, as settling, or tending to settle novel and intricate questions of law relating to patents, arising not merely in this case, but affecting many others. We will not linger upon this view of the subject, feeling confident that it must long since have attracted the attention of the Court. We feel sure that your Honor will consider, with anxious deliberation, questions which embrace such vast interests, and on the decision of which the business, the property and the pursuits of so many thousands of individuals depend.

What are the grounds upon which the Complainant asks for *your Honor's interposition?* As I understand the learned counsel on the opposite side, they say that the Plaintiff has a patent, and, also, that he has a valid extension thereof. That he has had for a long time uninterrupted enjoyment, or what he calls public acquiescence in his claims; that he has obtained two verdicts; and a judgment in this Court upon one of them; that the Defendants have infringed his patent, and are doing irreparable injury to the Plaintiff, by building cars and selling them to the Railroad Companies; and that he has not said or done any thing, which would render it inequitable to put a sudden stop to the business of these Defendants ere he can establish his rights before a jury, and thus pave the way to annihilate all the travel on the railroads of this busy community. Such are *his* assertions, such he contends are the facts ; and thereupon he demands *your Honor's intervention.*

I do not propose to spend a moment in stating elemental principles which regulate the issuing of injunctions, excepting so far as to call your mind back, for a single moment, to the reasons on which the power of granting them is founded, and to the just limits of the exercise of judicial discretion.

By the statute of 1789,—the Judiciary Act,—the power to grant *injunctions* is only where " *there is no plain, adequate and complete remedy at law.*" The Act of 1836,—the Patent Act,—authorizes injunctions *but according to the course and usages of Courts in Equity.* Not according to the course and usages of *English Courts* of Equity, *but according to the course and usages of the Courts of Equity as established by the Judiciary Act.* Now there ARE CASES of infringement in which a suit at law does not give a FULL and ADEQUATE REMEDY. As 1st, Where the Defendant is IRRESPONSIBLE ; 2d, Where his business is of a fugitive character ; 3d, Where he has had a trial, and being sued a second time, has discovered NO NEW DEFENCE ; for the Court will not waste its time in hearing the same thing over again ; and 4th, Where the *same case*, on the part of the Plaintiff, has been repeatedly sustained, and the *same defence* then set up has been repeatedly OVERTHROWN, WITHOUT COLLUSION, or the appearance of collusion, and when there has been a full and HONEST DEFENCE.

In SUCH CASES an injunction alone will protect the rights of the Complainant, because, otherwise, he might be forever compelled to litigate, and there might be no spot of ground on the face of the globe where he could rest his foot and find his title to that enclosure secured to him by the Courts of his country. BUT, an injunction is not *necessary*, nor within the true scope of the power of the Courts, where a suit at law is " *a plain, complete and adequate remedy.*" And an injunction IS NOT to be resorted to in such cases ; *e. g.*, where the Defendants *are* unquestionably responsible; where their business *is* local and permanent, there being no chance of escape from responding to judgments at law; and where the Plaintiff is not engaged in any line of business in which an injunction is necessary for his protection against competition. The character of our Government requires that every man should have his common law rights, and first among them his right of trial by jury preserved to him, unless some extraordinary circumstance warrants EQUITY in interfering and withholding them. To preserve these rights was one of the great objects of the Judiciary Act. Now, the granting or withholding of injunctions, is always discretionary. It is *never compulsory.* There is no state of facts that can be proved, in which the Court is absolutely bound to yield up its judicial discretion, no *matter* what are the merits of the inventor; no matter how many verdicts have been rendered in favor of the Plaintiff against strangers; nay, even against the same parties; the injunction is still held within the Judge's hand, and it may be allowed to go forth or be held back, according to the opinion of the Court, considering all the circumstances of the case. We mean to say that there is no imperative rule upon the subject, and we need not refer to the many cases disposed of, not only in this Court, but in others. We will suggest that of Maney vs. Sizer, in the District of Massachusetts; where, after long investigation by some of the most distinguished counsel in our neighborhood, a verdict was rendered sustaining the patent right, and finding that the Defendants had infringed that right, nevertheless, the Judges who presided at that trial refused to grant an injunction, under the circumstances, against even the Defendant himself. Numerous cases, to the same point, are cited in " Orr vs. Littlefield."

I also call the attention of your Honor to the decision of this Court in this case, in Wood and Minot's Reports, cited by the learned counsel for the Complainant. And in this very cause Judge Conkling has refused to grant an injunction notwithstanding what has transpired before this Court. We are free to admit, that injunctions are usually granted in the Courts of the United States, when the Plaintiff's right is WELL ESTABLISHED in the following manner.

First. *By unimpeachable verdicts, which have been rendered upon a full investigation of* ALL THE FACTS *brought to the notice of the Judge, before whom the motion is pending.* With your Honor's permission, I will repeat that statement. " By unimpeachable verdicts, which have been rendered upon a full investigation of ALL THE FACTS brought to the notice of the Judge, before whom the motion is pending." That is the qualification. That is, if a verdict had been rendered upon *all the facts* that we have *now* been spreading before your Honor, we should have comparatively little ground to object to the issuing of the injunction.

Second. By LONG and EXCLUSIVE and UNDISPUTED acknowledgment, on the part of the public, of *the rights of the Plaintiff as he claims to have them construed* in this Court. Not of SOME rights. Not of SOME claims. Not of SOME patent. But of the *very* rights which he claims your Honor's sanction to uphold.

Third. Where there are NO CONSIDERATIONS brought to the attention of the Court, that would make a sudden interruption of the Defendant's business unjust and inequitable.

Fourth. When the *injury* to the Defendant's business would not be irreparable (in case the Plaintiff should, at last, fail to maintain his claims).

And lastly. Where the *injunction* would be the *only* effectual means of *securing* the Plaintiff from IRREPARABLE LOSS. *Not* where he is SURE of OBTAINING ample satisfaction for the use of his invention, in case he succeed in establishing his *claims.*

These are familiar rules, and we cite no authority to support them. At the trial before Judge Conkling, it did *not* appear that there had been any thing approaching an *undisputed possession* of the rights claimed by the Complainant,—any substantial acknowledgment on the part of the public as to the validity of those rights; nor, *any such proceedings in the shape of verdicts* as would sanction the Court in using this tremendous power of injunction; and the attention of the learned counsel for the Plaintiff was called, we presume, to that fact, in the excellent opinion of his Honor. The Plaintiff has come forward a second time to build up, somewhat higher, his airy castle, and having produced some farther evidence in regard to public acquiescence,—although the testimony that was produced on the former trial was quite insufficient to move his Honor's mind,—it is my duty to present it to the attention of this Court, to state the evidence fairly and the answers that are to be made to it.

That evidence consists of two classes of proofs. First,—LICENSES. And, second,—SUITS. I speak, first, of the Licenses, or what may be treated as some evidence of acquiescence on the part of the Railroad Companies. The first mentioned is the transaction with the Baltimore and Ohio Railroad Company; and I need say no more than that a contract was entered into between that Company and the Plaintiff, *not relating to this patent*, and having no reference to it, but merely giving that road a title to the services of Mr. Winans in any improvements that he *might have made*, or *might thereafter make*, in relation to the machinery of the road. This patent was not at that time in existence, and, as it was no part of that contract,—or rather, as nothing in particular was said about it in that contract,—we may presume that this went into the general category of all other improvements that might thereafter be made; and it could not be treated as a license appertaining to this patent, nor an admission of the validity of that claim, *which did not*, at that time, exist.

Second. The Frenchtown and Philadelphia, and Wilmington and Baltimore Roads subsequently *united* into one road. The evidence on that point proves, that a suit was brought against the York and Frenchtown road, before his honor Chief Justice Taney, some time since. The Plaintiff in that case failed to obtain a verdict. After the suit had been thus suspended, and the Jury disagreed without acknowledging the Plaintiff's rights, the small sum of $500 was paid to the Complainant, *causa pacis* and to avoid all further litigation; a sum which I trust was far less than the fees of the learned counsel who managed that cause. I know not that anything need be added to these remarks upon that subject, except to refer the Court to the testimony of Shultz and Dorsey, who both say, that they *never did use a car substantially like* the Plaintiff's upon that road. Therefore, taking this fact in connection with this quiet and peaceable adjustment of litigation for the small sum of $500, it seems to me the circumstances are far from proving a quiet enjoyment of the rights claimed by the Plaintiff, or the receiving from him any license, or employing or using under such license, cars substantially like that which he claims. I think that this court will not look upon such adjustments as admissions of rights, and that it will not be ready to discourage peace-making by misconstruing the motives of the parties.

The third alleged license is that comprehended in a contract with the Boston and Lowell Railroad Company. But, on looking at that contract, which is annexed to the affidavit of Mr. Higginson, your Honor will find that it embraced the right to use four of Mr. Winans's patents, including this, and that the engineers of the road, and others acquainted with it,—as, for example, Mr. Higginson, Mr. Lee, Mr. Parrott, and Mr. Griggs,—all united in stating that the Boston and Lowell Railroad Company have never used a car constructed substantially like that of Mr. Winans. Mr. Higginson goes further, and says that of so little worth was that license, so far as it related to Winans's eight-wheel trucks, they would not have given any thing for the contract, if it had not embraced the other three patents, which were supposed, unhappily for them, to be of some value.

Again. The Reading Railroad Company is the last mentioned in the evidence to which my attention has been turned. And it is stated that Mr. Tucker, the President of that road, purchased of Mr. Winans the right to use his eight-wheel trucks under all the freight and passenger cars on that immense road, of ninety miles in extent, and doing a business of more millions of dollars per annum, than I dare undertake to remember. My colleague at this moment informs me that the sum annually earned is between two and three millions of dollars on freight, thus employing thousands of cars. For the paltry sum of $500 Mr. Winans sold the right to use his invention, if he had any, to that Company. Mr. Winans states in his bill, that one, whom we all know to be among the most distinguished members of the bar in the United States, I refer to Mr. Meredith, was his counsel. And it also appears in the evidence in this case, that Mr. Meredith was at the same time the counsel of the Reading Railroad Company, from which facts, together with the high character of that gentleman, both in and out of the profession, I infer that this paltry sum of $500 was rather paid as a compliment, and out of respect to Mr. Winans's counsel, than with any view of acknowledging Winans's claims, and not as a compromise. Had it been $50,000, it would be more like what Winans pretends that his patent is worth. And if your Honor will look at the grant for which $500 was given, you will see that a little *douceur* of that kind, obtained from the Company which Mr. Meredith represented, was a matter of courtesy or compliment, rather than to be considered as a formal acknowledgment of Mr. Winans's claims, be they what they may. And further, Mr. Tucker has sworn, in his affidavit, that *no such car* as Mr. Winans's either *now is*, or *has been used* upon the Reading Railroad ; notwithstanding the right of the Company to use it if they pleased.

Now if the Plaintiff RELIES upon his sale of *licenses* to a few Companies, as evidence of undisputed possession, or acquiescence on the part of the public, we state in reply three things : *First*, that of alleged licenses, considering the *very small* number, (being four only with the largest construction,) compared with the universal use of what the Plaintiff claims as his invention, *the fact* of there being any licenses existing, only causes the almost universal NON ACQUIESCENCE to appear the more boldly and clearly. And *Secondly*, that these licenses were for paltry sums of money, taken *causa pacis*, WITHOUT a pretence of their being adequate to any substantive right. And, *Thirdly*, that such licenses, unless numerous, are *not* evidence of much value, and unless acquiescence is general, it has no force in a Court of Justice. A man can always obtain acquiescence on the part of his friends, and unless it is universal or approaching to universality, the evidence is but of trifling import.

I now, with your Honor's leave, approach a more important matter. The Plaintiff relies upon former trials and verdicts, and a judgment upon one of them, doubtless, as putting the validity of his claims beyond dispute. There have been three trials in the Courts of the United States, and each of them has *some* importance. First, the trial in 1838 of Winans vs. The Newcastle and Frenchtown Railroad Company, before his Honor Chief Justice Taney. I must say of that trial, although it resulted in favor of the Defendants, that having examined the minutes, and particularly the opinion, the real facts, now disclosed, were then but partially known; and they were brought forward under every disadvantage, on the part of the defence; and notwithstanding this, the Jury *did not find a verdict for the Plaintiff!* The trial was a failure: it was too near home; too near the source of information in regard to the early history of the Baltimore and Ohio Railroad; too near to make it desirable to repeat the application for an injunction, in that part of the country.

Nine years after the first effort, in 1838, which was four years from the date of the patent, what does the Plaintiff? He waits in the first instance four years, before he gets defeated, and then goes off with the paltry peace-offering of which I have spoken. What does he next? Does he go forward and press for another trial against some other party? Does he seek another forum, and there endeavor to enforce his long deserted claims? I think, that large as your Honor's experience has been in patent cases, the memory of the Court will be taxed in vain to recall a single instance, in which an inventor who did not believe his patent worthless, should wait NINE long years before making his second appeal to the Courts, and SIXTEEN years before he obtains his first verdict. What would the country infer, if the trial before Chief Justice Taney had been made known? Would it not be said that there has been an effort to sustain these pretended claims before a Chief Justice of the United States, which had utterly failed? And when the Plaintiff, after one defeat, is unheard of in the Courts for nine long years,—I ask what inference a business man would justly draw from that fact? Such a defeat would not, as the Plaintiff alleges, tend to *make* every one settle with him, and admit his asserted rights, but it would lead them to suppose that he would cease from enforcing his patent in any other quarter.

Then Winans waits until 1847, in which year we hear of the suit of Winans against the Troy and Schenectady Railroad Company, and after that has been allowed to linger by the wayside until 1850,—sixteen years after the patent was originally granted, —he then obtained a verdict for $100! and had judgment upon the verdict! And now, I ask your most careful consideration of the *true weight to be attributed to that verdict,* in the decision of the present cause. I hope your Honor will not take me to be one, who desires to diminish in the slightest degree the solemnity and importance of a verdict rendered in this Court of the United States. I do not intend to utter a syllable against the verdict, nor to ask your Honor to undervalue, to the thousandth part of a drachm, the true weight to be given to it. And I know that your Honor will take care that the weight which does not belong to it shall not be given to it, in the disposition of the cause which we have the honor to present. If it had been a verdict like those in the ordinary course of trials, no objection to be made to it, and nothing to be said of it, except that it disposed of the whole case, I should not have come here to argue this cause. But allow me to call your attention to some of its peculiarities. And, in the first place ; it is not doing injustice to the shrewd management of the younger counsel of Mr. Winans, to give him the credit of saying, that this patent has been pushed forward, in these last days of its existence, with an adroitness which is worthy of a grey-headed counsellor. And the first step in such cases is, of course, to gain verdicts where they can be most easily obtained ; to attack Defendants at their weak points ; just as an invading army attacks another, namely, where they are most hopeful of victory, and where the least resistance is to be met with. Why did they not go to Baltimore and attack some of the Railroad Companies, which are so ably represented by the counsel for the Plaintiff? Why not go to Massachusetts ? They come to the State of New York, it is true, but they select their victim. It is the Troy and Schenectady Railroad Company. I am free to say that the cause was in the hands of the ablest men that this neighborhood affords; but we say that the road itself was so situated, that it had not the means or the disposition to go into a tedious and expensive investigation of the facts of this case. That is enough. A road which, at that time, was about expiring, and which has since been sold out for a small part of its original cost,—that road was attacked while struggling for existence and scarcely able to stand alone. Another feature of the trial which deserves consideration, is, that there was an agreement between counsel at that trial that the verdict, if rendered in favor of the Plaintiff, should be only for the nominal sum of $100. The Plaintiff stating, as of course, that the object of that suit was merely to establish the right, and not to injure the Defendants. And although such agreements are not unusual, and are made in good faith, I must say that they greatly facilitate the jury in passing to a verdict in favor of the Plaintiff, and that verdicts thus obtained are entitled to less weight than if rendered by an unbiassed jury, and intended to indemnify the Plaintiff for the damages actually sustained in the alleged infringements of his rights.

It was not a ground of defence in the Troy case, that *after* the date of the Plaintiff's patent he had acquiesced in the public use of his invention without objection or claim of right. For proof, I refer to the affidavit of one of the gentlemen who was of counsel in that case, Mr. Buel.

The defence rested, mainly, upon the testimony of a few experts, who supposed that the true legal construction of the claims of the patentee was such as to embrace every description of eight-wheel car which had two swivelling trucks under it. That as Chapman, Wood, and Jervis had invented and put into public use running gear which answered this description, these prior inventions were sufficient to defeat the patent.

These experts misunderstood the nature and extent of Winans's claims, as shown by the subsequent decisions of this Court; they, therefore, misapplied and misused the testimony contained in the books and drawings produced at the trial. For although the Chapman car might not be sufficient to defeat the Plaintiff's right to a patent, for the alteration he introduced, or the improvements he made upon that car, yet it might be and was sufficient to protect the Defendants in their right to use precisely what the Chapman car embodied.

The Chapman car, therefore, should have been put in proof, not to show that Winans was not the first inventor of what he claimed, but that the Defendants had used nothing in the construction of their cars, excepting what is found in Chapman.

This mistake of the experts rendered the evidence wholly worthless at the trial of the Troy case; and it is presented at this argument for purposes wholly different from those entertained before.

We assert, thirdly, that the principles of the eight-wheel car had not then been fully and scientifically investigated and practically tested by experts who were known to the Defendants.

These cars had gone into public use, without attracting particular attention, had never been the subject of that species of discriminating analytical investigation which is necessary to distinguish the mechanical principles of one peculiarity from another in the construction, organization, or modes of operation of different structures.

Fourth. That the *Wood and Trussel Cars* were not inquired of in that case. The Defendants in the Troy and Schenectady Railroad Company—what did they know about the history of *Wood and Trussel Cars* in Baltimore in 1830 ? Not a solitary sentence was uttered upon the subject, although the pay-rolls of the Baltimore and Ohio Railroad Company spoke of their manufacture and use ; and, therefore, we know that they existed. The Defendants did not know it.

Fifth. We now offer overwhelming testimony, that if the patent be so *construed* as to cover the Defendants' cars, that the *invention* was absolutely *abandoned* before the patent was taken out. Also, the conduct of Winans towards these Defendants would amount to a *license* to them to make use of the Complainant's invention. These considerations had great weight with Judge Conkling.

Sixth. The true use of evidence as to the old *Chapman Car* was wholly *misunderstood*, not only by the experts, as already shown, but by the Court and counsel. And inasmuch as the idea of the Chapman Car was not accurately comprehended, it was said that Chapman's invention was the same as that of Winans. We now say, that the Chapman Car is the same as ours, and that Winans's *differs* from *both*. In other words we say, that what distinguishes Winans's from Chapman's, also distinguishes Winans's from the Defendants'.

Seventh. A great body of evidence, of new and different cars, the same in principle as those of the Defendants, is *now brought to the attention of this Court for the first time*. I refer to the *Tredgold Car*, the *Allen Steam Carriage*, the *Trussel Cars*, the *Wood Cars*, the *Fairlamb Patent* and the *Quincy Cars*. Now, if these cars had been known at the trial before Chief Justice Taney, and his instructions to the jury had been applied to them, where would the Newcastle and Frenchtown case have anded? If these cars had been known to the counsel for the defence in the Troy case, how would that have resulted?

No person has more profound respect than myself for every considerate decision of an enlightened judicial tribunal. But I think that there is no class of subjects on which there is more danger that the judges may be uninformed and mistaken, than in questions of patent law, in its application to particular cases. The principles of the law are plain. But there is difficulty in the application of those principles to facts. And is it surprising that errors should sometimes occur in deciding cases, which require a court to dispose of profound and disputed questions, purely scientific in their character, in addition to the questions of law? I will ask your Honor, if it be not true, that this Court decides the construction of every patent according to the language of the patent itself;—and by aid of the light of all surrounding facts and circumstances proved to exist, at the date of the invention? If the Plaintiff were supposed by the Court to be an original inventor of the first eight-wheel double truck swivelling car that had ever been organized, or, in other words, that he was the inventor of the principle of the eight-wheel car, and your Honor was not informed that any other had been made before the Plaintiff's invention, then, another eight-wheel car is constructed, which is alleged to be an infringement of the first, would not you say at once, that the Plaintiff, being the first to discover and produce any double-truck swivelling car, is entitled to the principle as embodied in that structure; and that, as the Defendant embodies that principle in one form or another, he is an infringer? But how would your Honor's construction at once be changed, if it be proved that there existed twenty or thirty eight-wheel swivelling cars of various constructions, prior to the Plaintiff's invention? You would say, that every patentee's claim shall be so construed, "*ut magis valeat quam pereat.*" You would say, that if there be any construction of the Plaintiff's claim, coëxtensive with his invention, and which may be arrived at by fair interpretation of the language used by the claimant, so as not to affect his claim to novelty, the patent would be construed in such a manner as that the Plaintiff should go unscathed, if possible. His claim should cover his own invention, and no more. If a series of trials on the same patent should arise in this Court, every successive investigation bringing new and important facts to the notice of the jury, you would be compelled to draw the claim into more narrow limits, from time to time, upon taking into consideration those facts, which would render a broader construction fatal to the patentee ; and thus the construction of the same patent in different Courts, or in the same Court at different times, would seem to be irreconcilable. Such cases have already occurred in the Circuit Courts. And, therefore, while I entertain the most profound respect for former constructions by the judges, of the claims of the patentee, I say, that it is upon the facts before the Court, in each particular case, that that case is to be determined. When, therefore, a new set of facts arise, I do not presume that the Court will, necessarily, arrive at the same results as in the former case, as to the construction of the patentee's claim. It would be unjust to the Plaintiff and Defendant to insist upon drawing the same conclusion, no matter how different the premises.

And now, applying these remarks to the Troy and Schenectady case, what should have been the construction of the patent at that time, if the Court had known the existence prior to the Plaintiff's invention, of ten or twelve well-organized cars, very much like the Plaintiff's, and incorporating in their structure the general principles of the swivelling truck ? How different from *that* which the Court must *now* adopt in view of all the *new prior inventions ?* What injustice was done to the Defendant by their own experts, in their idea on the subject of infringement !

The next consideration in relation to this Troy case, is, that we have offered a large amount of evidence to show, that Winans was *not* the *inventor* of the arrangement of the wheels and running gear in the car *Columbus*, but Conduce Gatch *was*.

The Court will notice, that I do not say, that Winans was not the inventor of some peculiarities in the car; but what I say, is, that the running gear, the arrangement of the wheels in trucks, and their connection with the car body by a swivelling pin, was *not* first devised and first in use even on the Baltimore and Ohio Road, by Ross Winans, but this was first done at that place by Conduce Gatch.

In the Troy suits, the testimony as to the Baltimore cars on the part of the Defendants rested chiefly on Gatch; and the learned Court, in their opinion, seemed to think that it was quite incredible, that Gatch could have made the arrangement of the wheels; although he was at work on them every day of his life; that he required some authority to be given to him, in order to put his inventive faculties into legitimate action, or to place the wheels nearer together or further off, although very little alteration, if any, was made. But *now* we have some *twelve men*, surrounding and *supporting* him with a mass of testimony that I cannot suppose it probable, or even possible, that a jury could resist. We have *now* also *proved*, that the drawing attached to Mr. Winans's patent did *not* illustrate the specifications; and not only so, but it did not represent anything existing *at the date of his invention*. But it represented a car, built in 1837, and from which the drawings were taken. And I do not hesitate to say, that the ruling of his Honor Judge Conkling, in regard to the effect of that drawing, is erroneous. If my memory serves me, it was held, that that drawing had the same effect as if it had been a part of the original patent. But such is not the effect which the statute gives to a drawing that has been restored; it certainly does not give such an effect to a drawing, not restored according to the statute, and containing things which the patentee never discovered or invented. Again; it was there asserted, and very little proof, if any, was offered to the contrary, that the *smoked drawing*, on which the Plaintiff lays the *foundation of his claim*, was in fact a drawing that contained the running gear of the car Columbus, which was the only subject of controversy. And it is now, we think, clearly proved, that however much Winans might have done in planning, or furnishing plans of the various shapes of bodies, the plan of the running gear was *no part* of the *original* drawing. On this point the testimony is conflicting; but we are of opinion that any jury would decide with us. It is *now* proved that the Columbus, and the eight-wheel cars which preceded it, embodied all that the Plaintiff claims. It embodies more perhaps, and possibly it does not embody the exact details of structure, but it includes what his counsel claim as Winans's invention, as I understand them, although they now assert that the Columbus was an *experiment*, and that the experiment failed. Now it is proved, that eight-wheel cars, for different purposes, were in general use, both *before* and *after* the Plaintiff's patent. That fact was not dreamed of in the Troy case. Again ; it is *now* shown, that Winans's pretended invention, taking him at his word, is impracticable and useless, in so far as he departed from what was well-known before. Then it was *supposed* that Winans was the first man who had made any description of eight-wheel car, and that his claim covered every species of swivelling apparatus that could be imagined. That was the general idea at the time of that suit ; and in this mistaken idea, you find the reason why the Chapman car was made use of in the manner before-mentioned, and the idea unfortunately led to an entire misconception of the whole defence.

It is enough, in addition to what I have already said on this subject, to say that Judge Conkling refused the injunction twice, notwithstanding the verdict. And he knew well all these facts. This Court has pronounced its decisions by instalments, and Judge Conkling has twice decided these questions. I am aware, that the second time, the refusal was not absolute ; and he refused the injunction, even though there was an erroneous statement in Winans's bill, and one much relied upon by the Judge, namely, that he (Winans) had published his specification in the American Railroad Journal, and thus made the claim early and extensively known to the country. Subsequent investigation has shown this statement to be incorrect.

These are all the facts which I intend to bring to your attention, to show that this suit was commenced and prosecuted under peculiar circumstances ; that, except the evidence upon the Chapman car, (which we propose to discuss at the proper time,) no part of the defence now relied upon was produced or known to exist at the time when that case was tried and decided. And under these circumstances, this case stands before your Honor as favorably as though there had never been a trial. And we say further, that *this case*, and those facts now for the first time proved, *have never been passed upon* in this or in any other Court.

The third suit I have to mention is that of " Winans vs. The York and Maryland Line Railroad Company." Upon this point we shall detain you but one moment, and we beg leave to refer to the affidavit of Mr. Magraw, page 49, and more particularly to that of Mr. Robert S. Hollins, page 46, both in Defendants No. 2. Mr. Hollins states the facts under which that suit was commenced and prosecuted. There was no trial on the merits, and no investigation of the rights of either party. It was taken for granted, that Winans's patent was good, and that the cars complained of were built on Winans's plan, and reliance was placed on one great question, which has yet to be settled, namely, Whether or not the Railroad Company, which was the Defendant in that case, had used any cars, or whether such user was by another and different corporation ? And that question was decided in favor of the Plaintiff, much to the astonishment of the Defendant. I do not mean to assert that it was not rightly decided, but my object is to show your Honor, that no jury has pronounced upon the questions now submitted to this Court. And the reason why the Plaintiff's pretension and the alleged infringement were not disputed, is explained by the affidavits of these gentlemen. They placed their reliance upon the fact that they had not used the cars. That case, however, has not yet been finally determined ; it is pending in the Supreme Court, there having been no judgment; therefore this verdict cannot affect your Honor's mind. The use made of a verdict is, to inform the Court whether the allegations of the parties are true, or not true—whether the defences relied upon have been subjected to the examination and opinion of a jury ; and of course, as that cause now stands, you cannot be thus informed. Again ; and in the same case, the Plaintiffs have avoided bringing suit where he resided, and where the Company held its office ; but he has caught the President of that Road on a visit to Philadelphia, and there brought the action. And yet the use of the patent right was offered by Winans to that Company for $500, which sum they refused to pay. There is, therefore, no intention to admit, and no real admission is to be inferred from that case. Notwithstanding that this trial was before Judge Conkling, he refused the injunction. That, I consider as an opinion given by this Court upon the insignificance of the case, in relation to the question of granting the injunction now prayed for. In not one of these suits was the main bulk of our testimony brought to the notice of the Courts, and only part of it was offered before Judge Conkling, at the hearing when he refused to enjoin, and we have added great strength to our defence since that refusal. And, if any thing is to be considered in the course of this trial by this Court as settled, it is this: that a motion for an injunction cannot be granted, unless the complainant makes out a much better case than when the former injunctions were refused. *This* is the *third* motion for injunction! and I think it shows the great liberality of Courts of Equity, that a third attempt to obtain an injunction is for a moment tolerated.

I have now done with the *verdicts* and the *licenses*; and with all that which the Plaintiff sets up as furnishing any evidence of acquiescence on the part of the public, or as tending to substantiate his claims by aid of any judicial proceedings.

I now propose to show that the Plaintiff ought *not* to have an injunction under the peculiar circumstances of his case, and of our position. The Plaintiff is ESTOPPED from *relief in equity* by his acquiescence in, and encouragement of, the use of the invention (if it ever was used) by the *Defendants* and by the *public*. I will not take time in reading authorities, but I beg to cite three or four, and then leave this subject. I refer to Curtis on Patents, p. 395, note 2;—Wyeth and Stone, 1 Story, 273;—Neill vs. Thompson;—Webster's Patent Cases, 275;—same case in Hindmarch, 319, 20;—Saunders vs. Smith, 3d Mylne and Craig, 711. Also see pp. 720 and 736, Greenhalgh vs. Manchester;—same book, p. 785. And I further add upon that point, that this is *true* even though such acquiescence should not go so far as to constitute an *abandonment*, in law, of the Plaintiff's rights. For *equity will not tolerate a trick, a fraud, a surprise*. Winans well knew that the Defendants were engaged in *making and vending* eight-wheel cars for years. He made *no objection;* he *gave no notice*. He *assented impliedly;* and even joined the Defendants in sending eight-wheel double truck cars, such as he now claims are in violation of his patent, to Germany. He united with my client here, (Mr. Orsamus Eaton), in sending just such cars to Germany; but he never lisped a word about any patent, nor objected to it, as in any manner infringing upon his alleged invention. Again. It is shown that Winans has been selling car wheels, out of his own establishment, to Eaton, Gilbert & Co., well knowing that these wheels were to be used in making the cars commonly built in this country, and which he here claims as violating his patent.

These three things : *uniting* with the Defendants in sending cars to Germany ; *selling wheels to them* for the purpose of *being used* under those cars ; and having numerous conversations with them, *without* alluding to this *subject*,—are, what any man could not help calling a *consent* on the part of the Plaintiff to what was going on. It would be a fraud by Mr. Winans to set up a claim to damages for an infringement, after such conduct as that. Again. Eight-wheel cars are known by everybody to have been in use on all the Railways in the United States. And the Plaintiff knew it, and yet made no objection to it until recently. Up to 1847, thirteen years after the date of the patent, and until one year of the end of its natural existence, nobody ever heard of a claim to such an invention. It is not until after its death and resurrection, that it assumes a menacing attitude. I will not go further than to say that I hold before me the list of names of twenty-one witnesses, who show that Winans well knew that the eight-wheel car was in universal use ; that he made no objection to such use, and no claim to it; so that he must have known that such cars were no violation of his patent, or that his patent was good for nothing. And I will simply refer the Court to the name and the page where their testimony is recorded. See affidavits of

1. John Murphy,	Defendants',	No. 3,	p.	9
2. Laban B. Proctor,	"	" 3,		8
2. Wm. Raymond Lee,	"	" 1,		41
4. Charles Minot,	"	" 3,		159
5. John Wilkinson,	"	" 2,		42
6. Robert Higher,	"	" 3,		132
7. Jeremiah Van Rensaellaer,	"	" 4,		44
do.	"	" 3,		136
8. Albert Bridges,	"	" 1,		37
do.	"	" 3,		134

22   INJUNCTION.

    9. Charles Davenport,   Defendants', No. 3, p. 39
  10. D. N. Pickering,   "   " 1,   39
  11. George S. Griggs,   "   " 1,   40
  12. Timothy S. Smith,   "   " 1,   37
  13. James L. Morris,   "   " 1,   39
  14. Edward Martin,   "   " 1,   45
  15. John Stephenson,   "   " 1,   42-3
  16. Leonard R. Sargent,   "   " 1,   43-4
  17. George Law,   "   " 2,   29
  18. David Beggs,   "   " 2,   31
  19. David Mathew,   "   " 2,   32
  20. Israel Adams,   "   " 3,   138
  21. George Beach,   "   " 2,   136

Now, that is the list. There are twenty-one of them; and without turning to the evidence, I will simply say, that they comprise prominent men connected with some of the most important railroads in the Union.

Now, what is the rule of Equity and Law, with regard to acquiescence? The *rule of law* is, that where an owner stands by and sees others sell or use his property *without objection*, he *waives his right to claim it*. If a man sell my horse in my presence, and I do not object to it, I cannot afterwards set up my title as against the vendee. And the law applies with equal force to patent rights, as to any other description of property; and it appears to me with justice—because a patent right is in some sense a species of intellectual, not a tangible property; more liable to be appropriated to the use of others, without intentional wrong, than any other species of property. Ideas have no " ear mark " by which they can be traced to their owners, and therefore the greater good faith is required by all patentees, that they do not so conduct themselves as to mislead others by acquiescence in the public use of their inventions.

I next pass to the consideration of the evidence as to the notice which is said to have been given by the Plaintiff to Defendants. As to this, there are two witnesses on the part of the Complainant, who assert that a notice was given to one of the Respondents, the Plaintiff himself, and Mr. Gould who is a party in interest.

The counsel on the opposite side, say that Mr. Winans had a conversation with one of the firm of Eaton, Gilbert & Co., and that he made some remarks about his patent. And Mr. Gould asserts distinctly, that *after* the Troy suit, he gave some notice to the Defendants. These are the two gentlemen by whom alone the assertion of any notice whatever, is made. Now, Eaton, Gilbert & Co., all three of them, come into Court and ANSWER under oath, and positively and unequivocally *deny* that they received notice of any description from Winans or Gould. And, *on the contrary*, they state, that they *never received* such a notice.

I am speaking of notice asserted since the Troy suit. Here is a direct contradiction on this question. Three on one side and two on the other, and I do not wish to suppose that any one of the gentlemen intends to state that which he does not suppose to be correct. But I do suppose that Mr. Gould, having a large area to travel over, and a vast number of persons to see, might have *confounded one* of the firm of Eaton, Gilbert & Co. with *some one else*. And in regard to Mr. Winans, he may have *supposed* that he *might have* said something about his patent. Therefore, as it stands upon the affirmation of the Plaintiff, and denied by the Defendants, it is not proved. And Mr. Gould, although not a party in the suit, is a party in interest, and he ought to have been made a joint Plaintiff. It is proved by Minot' that he is a party in interest. Thus we have one, speaking of a conversation,—if it ever did take place,—and another asserting that a remark was made in the presence of one of the Defendants, and both of which assertions are denied by them in toto. I will now pass to the notice which it is *alleged* has been given to the *public*. The Plaintiff *asserts* that he published his specification in the American Railroad Journal. If that be true, it would be giving those parties mostly interested in railroads a notice of his claim, and a man could not be expected to do more. But the fact is, that he did *not* give that notice, and I refer to the affidavit of Mr. Randall, and to the fact that we have given notice to produce that Journal, and it has been produced and the notice does *not* exist. Therefore it is a *mistake* of Mr. Winans, and he has been led into some error, I know not how; but, of course, no man is free from mistakes.

Then, again. It is alleged that the extension is relied upon, and that there was a publication then. All I have to say on that point is, to ask your Honor to look at the papers for yourself, and you will find that it is not so. There was no insertion in the " True Sun," in which it was ordered to be published. But it was substantially concealed, by placing it in a small ephemeral paper not *generally* circulated, and, therefore, we did not get the notice that we ought to have had. For, if it had been known here, it is not to be supposed or imagined but that there would have been an opposition to the extension. At all events, the Patent Office would have had to examine the evidence that has been submitted to your Honor. I will briefly refer to the statement of Mr. Winans, to show that he has made four mistakes, and that perhaps his memory is not so reliable now as when he was a younger man.

First. I refer to the publication in the Railroad Journal, three or four times sworn to, and erroneous in point of fact. Second. To the opposition of the extension of his patent being on behalf of the New England Roads. On this point I refer to the affidavit of Lee. Defendants, No. 3, p. 110; to the record the gentlemen produced; and to Wilkinson's affidavit, No. 3, p. 50; and your Honor will see how incorrect his statements are. Then, thirdly. To the fact, that the Boston and Providence Railroad Company are said, *by Winans*, to have given an order for building cars on his plan, which is erroneous. See affidavit of Lee. No. 3, p. 110, and also to that of George S. Griggs. No. 3, p. 110. And the last is the one in controversy, namely; the conversation in which it is alleged that Mr. Winans said that he had a patent, and intended to enforce his rights. And, therefore, I argue that as Mr. Winans is mistaken as to documentary evidence, the probability is that he is also mistaken on this point. I am not aware that any intimation was given to any body of the existence of the patent before the Troy and Schenectady suit, while the patent had been laying dormant for so many years. And hence the palpable injustice to the railroads who would be prejudiced by a sudden injunction, after being lulled into security and led into danger by the Plaintiff's own conduct, as the roads have been led to *buy and use* these trucks, under the belief that they were common property, neither Winans nor any one else appearing to claim the invention.

Then, I ask your Honor to consider another reason for refusing the injunction, whatever the merits of the Plaintiff may be. It is, that the Plaintiff's LACHES in enforcing his claims is enough to prevent the Court from enjoining the Defendants at this late day. If the Plaintiff could wait from 1834 to 1852, without an injunction, there is no more probability of any injury being done to him now, and for the next three years, than during any *one* of the past *nineteen years*, when his own conduct shows that the Plaintiff did not consider that any *irreparable* injury had been done. He is now, as he has heretofore been, a manufacturer of steam engines and machinery. He is not in the same line of business as the Defendants, and, therefore, an injunction is not wanted to prevent competition or injury to the Plaintiff's manufactory.

Furthermore. The Plaintiff has never had any *exclusive use of the patented invention*, as he now claims it. His rights have never been admitted,—except once or twice,—*causa pacis*. He never asserted and followed up the claim which his counsel now set forth. And I ask your Honor, what *irreparable* injury will be done to Mr. Winans, by refusing an injunction until his rights shall be settled on final hearing, either in this Court or in Massachusetts? In the case to which I have referred in Massachusetts, the Court have ordered that the evidence shall be in in October, and it is our intention to try it as soon as possible. But no irreparable injury will follow before the case can be decided in one Court for the whole country. Another reason for denying the injunction is, that the Plaintiff has *no claim* whatever, to cover by his *patent right*, the cars made by the Defendants. We *deny* the *validity* of the *original patent*, if such be his claim, and the legality of the extension. And we deny, also, the *alleged infringement*.

Of these subjects I shall speak at another time. But the principle of law is, that an injunction shall go out only to protect unquestionable rights. *Not* to give one party an *advantage* before the question of right is settled. On this subject I have but a further word or two to say, and I ask your Honor's most serious and deliberate attention to it. By examining the affidavits of the Defendants, and their answers, you will see that the injury *to them*, in case this injunction be granted, will be irreparable. They have a capital of over $80,000 invested in the manufacture of these eight-wheel cars. They employ a large number of workmen, who, with their families, are dependent upon them for their daily bread, and who I believe cannot be employed by them, unless in the manufacture of these articles. The Defendants themselves would be broken up in business, for they tell your Honor upon oath, that they cannot apply their capital, or their structures, or employ their hands, excepting in the pursuits to which they have been heretofore devoted. I say nothing of the painful feelings with which they are brought here as trespassers, because your Honor will be influenced by no feelings of sympathy, however well deserved. But I *will* ask you to consider the statements in their affidavit, made without flourish and in all simplicity, and you will perceive the disastrous consequences to themselves and their business.

You will also see that these gentlemen are men of property, and of undoubted responsibility and integrity. They stand at the head of the car building establishments in the country, and are able and ready and willing, instantaneously, to respond to any just claim of Mr. Winans, in dollars and cents, if they have, in fact, in any way invaded his rights. And as the case in Massachusetts is so near its approach, it seems unnecessary, for the protection of Mr. Winans, under any circumstances, to inflict such a heavy blow upon these Defendants.

An injunction is sought here, nominally, against the car builders; but this is really an attempt to get control of all the railroads in the country. And if this injunction be granted, your Honor's Court will be thronged with applications for injunctions in order to compel parties to pay what is demanded of them, whether it be *right* or *wrong*. To suspend the business of a railroad for a single month might be utter ruin to the Company, and it would better submit to the most exorbitant demands, rather than stop the business on their road. Such power cannot properly be granted to the parties interested in this speculation in the exercise of *sound discretion*, whatever be the supposed merits of the Plaintiff's claim. To confirm these views, I will here refer your Honor to Judge Grier's opinion in the case of Oliver H. P. Parker vs. Sears et al., in which he absolutely refused an injunction, notwithstanding the Plaintiff had previously recovered repeated verdicts in different Courts of the United States; because he considered that the evil effects of an injunction would not end with the Defendants, but that they would involve other interests and other parties, and put a stop to their employment. It is an opinion filled with sound views on the subject of granting injunctions; wholly applicable to this case, and sustains the positions assumed in what has been already submitted.

In conclusion, upon this branch of the subject, we cannot but think it the greatest injustice to the Defendants suddenly to break up their business, and to throw their workmen out of employ, under the circumstances of the case.

It would be unjust to the Railroad Companies, to put a stop to their income from public travel; or subject them to the grasping cupidity of speculators in exhumed patent rights.

It would be unjust to the travelling public, whose business and pleasures would most unnecessarily be interfered with, by depriving them of the ordinary means of conveyance.

The builders of these cars, and the Companies who have invested money in their purchase, have made these investments innocently, honestly, and without intending to wrong any one. Such investments should be protected in Courts of Equity.

EXTENSION.

THE Defendants deny, that the *extension of this patent is valid.*

First. Because the *Commissioner failed to order such notice* as was required by the statute, which alone gave him jurisdiction of the case.

The notice actually ordered, is found in Wilkinson's affidavit, No. 3, p. 50. The statute passed May 27, 1848, about one week before the Plaintiff's application—required sixty days notice to be given. The former statute of July 4, 1836, required only three weeks notice. The notice ordered required three weeks, *ending* sixty days before the hearing. See Wilkinson, *ut supra*. Is this order sufficient, or does the law require the Commissioner to order a notice to be published sixty days, instead of three weeks ending sixty days before the time of the hearing?

Second. The *notice which was ordered to be given was not given in fact.* See Wilkinson's affidavit, *ut supra*, for the facts.

Third. The statute regulations for the extension of a patent, must be strictly followed. The notice was, in effect, concealed from New York and Massachusetts. The evidence actually existing, as to prior inventions, was not made known at the Patent Office. See the Record, which shows that fact. All the evidence, including the Fairlamb Patent, (which we suppose was then mislaid,) was unknown to the officer who extended that patent. The extension was opposed, only upon a very limited ground, without investigation of merits or knowledge of facts; and no notice was received, by those most interested to oppose it. The ground of objection taken, was, that the chief clerk had no right to make the extension. See affidavits of Lee, and Wilkinson, p. 54, No. 3.

Fourth Query. Is this matter *discretionary*, as to *what* notice is to be given? If it be so, and the Commissioner orders a certain notice which is *not given*, has he any more power?

Suppose that only ONE paper actually published the notice, could he pronounce that *sufficient*, after having required and ordered notice to be given in eight or ten papers? Suppose *no* notice is published: Can he dispense with it? Is not the compliance with an order of notice, as necessary to give him jurisdiction, as the compliance with an order of a Court of justice? Does the granting of the EXTENSION foreclose all inquiry? Is that CONCLUSIVE? Cannot the question of jurisdiction be investigated? Supposing that no notice had been given, it is clear that the Commissioner has no right to extend the patent. It may be said, that the statute is directory; yet it is nevertheless obligatory. It may be said also, that it is no fault of the patentee, that the notice ordered by the officer is not given. Nor is it the fault of the Plaintiff in a suit at law, that the sheriff fails to serve a notice ordered by the Court,—yet the jurisdiction fails.

Again. The patent was not extended by THE COMMISSIONER,

but by a CLERK, *who had no right to make the* EXTENSION. The Commissioner's authority is under statute. His duty in respect to extensions is *quasi* judicial.

Powers of this description cannot be delegated by a party to a substitute, though he be appointed by the government. Last winter an attempt was made to pass a statute remedying this defect. But it did not succeed. The effort clearly shows the feeling which was entertained by gentlemen on the subject. The question of the extension is very different from that of the ORIGINAL GRANT. The public have an *interest* and *right*, at the end of fourteen years, in the invention of the patentee. And the invention once having become public, or likely to become public, when the extension is asked for, the community have a vested right, and that is to be divested *only* by the exercise of a judicial act. The *origin and history* of the formation of the *Board*, to whom the duty of adjudging upon the extension of patents was assigned, shows the *importance* attributed to the exercise of their functions; and it is obvious that *under-clerks* were not in contemplation of Congress, as proper members of the Board. The only authority of the chief clerk to do anything is under the *Act of* 1836, section 2. See Curtis, p. 471. At the time that statute was passed, the powers of extending patents was invested in a separate independent tribunal: the Secretary of State, the Attorney of the United States, and the United States Commissioner. Their united action was necessary. The only power that the clerk had, was to do the *duties* which belonged *solely* to the *Commissioner*. The statute of 1836, which was sufficient to authorize the chief clerk to do *all* the Commissioner could *then* do, is not to be *construed* as extending to *all duties* that might thereafter be devolved upon the Commissioner. This is the nucleus of the whole argument. Powers, duties and capacities *subsequently* conferred upon the Commissioner, could not be transferred to his substitute, when those powers and capacities did not exist at the time when that substitute's authority was limited. The statute of *March* 3, 1837, section 2, provides that " copies of records certified by the Commissioner, and, in his absence, by the *chief clerk*, should be prima facie evidence," &c. That provision of the statute indicates that it was understood that whatever powers or duties were imposed upon the Commissioner, by statutes *subsequent* to 1836, were not imposed upon the *chief clerk*, or, without *special authority*, conferred upon him, even in so slight a matter as certifying copies out of the office ; and even though there were no words of restriction or limitation in the statute. I mean by that, that although the statute did not limit the power of making these copies to the Commissioner by any phraseology, yet it expressly conferred that power on the chief clerk, as well as upon the Commissioner. The argument *ab inconvenienti* on this subject is undoubtedly a strong one against us. For it would be sometimes attended with great injustice to a party, if an extension were defeated, because the chief clerk could not do any thing in the absence of the Commissioner. I admit all its force, and I think that the case ought to be provided for by statute. But that fact will not make the Court construe a statute more widely than the terms would warrant.

It has been decided, that for all duties devolving on the Commissioner alone, previously to the Act of 1836, the clerk was the Commissioner *pro hac vice*, in case of absence ; and also, that the Commissioner could take up an application for extension, *pending* before the Board, and go on and *grant it*. These two points are decided and settled by the cases cited on the other side. But neither of those authorities touch this question. And I have only to add, that as this is a matter of great moment, and as it lies at the foundation of this case, and as it is now in a situation to be decided by your Honor, in connection with your brethren at Washington, during the coming session, I will not go into the argument at large upon this question, but leave it here.

CONSTRUCTION.

IF the Plaintiff's claim be so construed as to include the cars of the Defendants, then we deny the *validity* of the original patent. But in order to render the objections to such a construction of the Plaintiff's claim intelligible, it is necessary to examine into *the nature and extent of those claims*. In other words, to put some *legal construction* upon the terms used in the letters

CONSTRUCTION. 31

patent. The subject, therefore, of the remarks I am now proceeding to, is, *the legal construction of the Plaintiff's claims*. And that construction may be ascertained by considering three questions or points.

First. The admissions of the Patentee, as derived from an analysis of the patent itself.

Second. By examining the language of the claims and the principles of Patent law, as to what *is*, and what *is not*, patentable, or capable of being embraced within the terms used.

Third. By ascertaining the state of the art of building the running gear for eight-wheel cars at the date of the Plaintiff's *alleged* invention, so as to determine in what sense he must be construed, to have used his language of description and of claim.

First. The admissions by the Patentee in his patent, are these :—1. That he was *not* the first inventor of *any one* of the parts that compose the car.

2. Nor of any particular form, or shape, or length of car body.

3. Of no mode of draft—none being stated, and none claimed.

4. That the Plaintiff's invention is for a FREIGHT or burden car, as well as for a passenger car.

5. Not for any exclusive right to place two swivelling trucks of four wheels each, under a long bodied car.

6. No particular size of wheels is recommended, and no statement made whether the wheels revolve *with*, or *on* the axes.

7. No *particular* distance apart of the bearing points of the wheels on the rails.

8. No particular distance of flanges ; by particular, I mean exact, excepting that they must be *very* close together—" *as near as may be without touching.*"

9. There is a statement of the *peculiar* theory upon which his whole philosophy rests,—of which I shall presently make an exposition.

10. No *proportions* of parts to each other given.

11. That Winans was *not* the first inventor of eight-wheel cars.

12. That bearing carriages were in *common use*,—which we

call " trucks "—and, that these would answer the Plaintiff's purpose, provided that "*the wheels were close together.*" See the Patent, p. 114, in Plaintiff's proofs, fol. 14. Nothing being claimed as new about *those* trucks, excepting the *approximation of the wheels.*

13. And, lastly, that there *were* eight-wheel swivelling double-truck cars, before his invention was made. Such are the admissions found in the patent *itself.* And I refer to Chief Justice Taney's opinion to show, that his Honor understood, precisely as I understand, these admissions of the Patentee.

What are the peculiarities of the car, as Winans would have it constructed? Let us examine them AS HE STATES THEM. They are,

1. Connecting the axles of the wheels in each truck, by a long spring, bolted to the boxes, twice as strong as those in the four-wheel cars.

2. Placing the wheels of each truck very near together, so that the two shall act, as nearly as may be, *like* a single wheel; the two trucks being at or near the end of the body.

3. Connecting the trucks with the body, by means of bolsters swivelling on a king-bolt, in the manner of a *common road wagon*, dispensing with any of the *side bearings*. And, your Honor will please notice:—And " bearing *the load on the centre* of the bolster," this is the language of the patent, " which is also the centre of each bearing carriage." See the Patent, folio 23, p. 5, Plaintiff's copy. " This arrangement", (he says,) " of the bolster, and the connection of it with the body in this peculiar way, affords *great relief* (Folios 15 and 16,) from *shocks*, occasioned by percussions of the wheels on protuberant parts of the rails, or other objects; and *from the vibrations* consequent upon the use of coned wheels, as the lateral and vertical movements of the body of the car, resulting from the above causes, are much diminished." Plaintiff's proofs, p. 114, folios 15 and 16.

He accomplishes the object of running smoothly and evenly over the road, freeing the car from vibrations, and giving great relief from shocks, by this peculiar arrangement of *resting the weight of the body upon the centre of the bolsters, without side bearings, and the weight of the centre of the bolsters upon the centre of the truck*; while *others* in *former times*, and *we* at the *present time*, adopt a very different plan, of which I shall hereafter speak.

Before critically examining the language of the Plaintiff's claims, let us see and understand what the Plaintiff's experts suppose that *his invention consisted in*. There are but *two* witnesses of any importance on this point. The first is W. C. Hibbard; p. 81, folio 384: and in a moment I will state the substance of his evidence. He says, that the eight-wheel car has three advantages over the four-wheel car; pp. 77-8. *Second*, that there are three essential points in the construction of the eight-wheel car; p. 79, folios 370, 371. Then he proceeds to distinguish Winans's from preceding inventions, by showing that it has three peculiarities; p. 81, folio 384, and which I will state. The *First* is, " the closeness of the wheels to avoid friction." *Second*, " to support the body on two swivelling trucks." And, *Third*, the application of the motive power to *the body*.

This is what he says *Mr. Winans's invention consisted in;* and it is quite important that the attention of the Court should be addressed, to what witnesses supposed the inventor invented; so that your Honor should see what their views are in speaking, both on the question of *novelty* and on the question of *infringement*. The other is John Elgar. P. 25, folios 106—10 of Plaintiff's proofs. He also gives his views of what the Plaintiff invented. 1st. He says, that he *thinks* the invention was a long body. 2d. Swivelling trucks at or near the end; and 3d. The wheels of the truck *closer* than would do in a four-wheel car. These are the Plaintiff's experts' ideas of what Winans *invented*. On what ground do his counsel place his claims to invention? The learned counsel who opened the argument stated, that " the complete principle of Winans's invention required 1st. ' The organization to have two four-wheel trucks;' and 2d. ' *Swivelling* freely under the body, by reason of transmitting the draft by the body, and the near proximity of the wheels in the truck, and the separation of the trucks from each other.'" So, that, if all *other things* are used by the Defendants, except drawing by THE BODY, they do not infringe!!

What I consider important in this admission is, that if there were "the near proximity of the wheels, and the remote position of the trucks," and *every other* feature stated in *Winans's patent, excepting the draft by the body*, that the invention was worthless, and those who used it would not infringe Winans's claim. If Mr. Keller is right in his view of what Winans invented, the claim would then be for a *combination* between an eight-wheel double truck swivelling car, and some apparatus for drawing it by the body. And the counsel do claim in *reality*, (though not in terms,) a combination between some mode of drawing by the body, and an eight-wheel car, which car, he says we may use *without* infringement. Because, if we may use the car without infringing, until we take that feature of the invention, then it is clear, that what he, in reality, means to claim, is a combination between some *new* mode of draft, and an *old* eight-wheel car. After having stated this much of what the *counsel* and the *experts* consider Mr. Winans's invention, I will now come to the language of the patent itself, and see if we cannot obtain some clear and definite idea of what *that patent is for*. He says:— and I read the language of the patent—" I do NOT claim running cars on eight *wheels*, this having previously been done." Not as he says in *the manner*, or *for the purposes* herein described. But *merely*, with a view of distributing the weight carried, more evenly upon a road, *and* for objects *distinct* in CHARACTER from those which I have had in view, as *herein set forth*." The objects set forth in the patent are these.

" The OBJECT of my invention is, among other things, to make such an ADJUSTMENT of the WHEELS AND AXLES as shall cause the BODY of the car to pursue a more smooth, even, direct and safe course " than it does as cars are ordinarily built, that is, on four wheels. As stated in this manner his object is to do certain things. The statement is not exactly correct, for he has confused the structure he describes, with the end or object for which the structure is designed. But the object is clear, plain, definite and distinct. It is to make the car travel more *smoothly*, directly— (not wabbling)—and safely. And there are no *other* OBJECTS stated in his *patent*. Various *means*, however, of effecting this result are stated. Now Winans denies in his patent, that railroad cars or carriages having eight wheels, were ever before arranged with the *same object*. The object is, "to run smoothly and evenly upon the straight and curved parts of the road." That statement might be accounted for, and is accounted for, if Winans was *not* aware, that there had been any preceding double truck eight-wheel cars, and he only knew of four-wheel cars being used. For, in comparison with the four-wheel cars the remark is just; but in comparison with the eight-wheel cars, the remark is most mistaken. I ask for what objects were the eight wheels of the Allen Engines, the Trussel, the Wood, and various other cars constructed; and for what object were they made to swivel at all, unless it was to run smoothly over the inequalities of the road, and go round the curves? But, whether these cars were built with that object in the mind of the constructor or *not*, still, if they were *calculated* to OBTAIN THAT OBJECT, it is not *material*. If the arrangement or *invention* EXISTED, it is no matter what was the *object* of the inventor. One who discovers the value of a PRE-CEDING INVENTION, is not thereby entitled to the *exclusive* use of that invention. Besides, if the eight-wheel double truck car had been originally constructed *merely* for the purpose of equally distributing the weight on the rails; and if it was *also found* that it would *run* smoother than the four-wheel car, that discovery is not such as to entitle the discoverer to a patent. For what object, or upon what design, were the eight wheels put in *two trucks* of four wheels each, other than to turn curves, and run smoothly? Eight wheels in a common stiff frame, would distribute the weight, under certain circumstances. But these eight wheels were put in two trucks; and the two trucks were to swivel, and turn curves, and run smoothly undoubtedly.

The claim goes on to say, "Nor have the wheels," when thus increased in number, "been *so arranged* and *connected with each other*"—either by DESIGN or ACCIDENT—as to accomplish this purpose. What I claim therefore as *my invention* is, the *before*-DESCRIBED *manner of arranging and connecting the eight wheels*, SO as to accomplish the end proposed by the MEANS SET FORTH, or ANY OTHERS which are ANALOGOUS and dependent on the SAME PRINCIPLES." If Winans had simply claimed the *before described* manner of arranging and connecting the eight wheels, this would have been *intelligible* to some extent; but this is NOT

HIS CLAIM. He does not *confine* his claims to THAT MANNER, or the means set forth, but he claims ALL OTHER means, modes, or manners, (*besides* those set forth,) which are analogous, *and dependent on the same principles*. So that he claims every *mode* there *can be, of making a car run more smoothly and safely than a four-wheel car, by any arrangement and* CONNECTION OF THE WHEELS, even though NOT DESCRIBED, or *invented*, or *thought of* BY HIMSELF, where it comes within a peculiar principle or theory which he sets out. Now, as the Plaintiff does *not confine* himself in his claim to the MEANS—and the machinery set forth—and to that which is *substantially* the *same*, but to ANY mode of accomplishing THE OBJECT, by *other* means than what he has discovered, it is *too broad* a claim. It is an attempt to get a PATENT for a RESULT—not a machine—or, rather, for *every mode* of arranging the wheels by which a certain *result* may be accomplished. If Winans means to confine himself to the arrangement and connection of the parts he has particularly *described*, THERE IS NO VIOLATION PRETENDED.

As the *patent cannot be* for a RESULT; that is, the result of running smoothly and swivelling to the curves—*not* being for any of the PARTS of the MACHINE—*not being* for the *trucks* alone, or *the mode of making them; not being* for the principle of the two trucks swivelling under one body, it is obvious that it CANNOT be for *merely* the *position of the trucks*, in relation to their distance from the ends of the body, as no *change* of position developes any *new* physical mechanical principle.

The patent cannot be for the mere *distance* of the *wheels* in each *truck*, as changes of distance develop no *new* mechanical principle; and the old statute declares, "that the *change of proportion* CANNOT *be the subject of a patent.*" Any one has as good a right as Winans has, to use the old elements in the same substantial construction, with a change of proportion of distances between THEM. *Change* of PROPORTION is a property incidental to all machinery, and, as I have said, not patentable. Another reason why the Plaintiff *cannot claim* the *nearness of wheels* in each truck is, because *he says*, that the trucks *in common use would answer his purpose, provided the wheels were close together.* It is proved, that the common bearing carriages had wheels from *three to twelve* inches apart; and that is far nearer than they are now put. So that, it must be supposed that Winans meant to claim, the putting them less than three inches apart, if distance has any thing to do with his claim.

Now, if his patent be for coupling the wheels in each truck, so *that the flanges shall come close together*, and *the two trucks far apart*, without reference to the peculiar mechanism by which this is accomplished, then it is *void* for *four* reasons.

First. That would be a mere change of proportion.

Second. It would not embody the true principles of an eight-wheel car, but the departure from those principles would be a pernicious change.

Third. The true principle of the eight-wheel-car was known *before*, and is used by the Defendants now.

Fourth. This arrangement as to distance, separately considered, is *not original with Winans*.

What is Winans's theory? What is his philosophy? What is the *basis*, the *radical idea* of his invention? It is this. To bring the axles of the wheels, as near as possible, to coincide with the radial line of the curves of the road, so as to approximate to the action of a single wheel. That is his theory, so far as regards the arrangement of the wheels; and, it is highly important that it should be clearly comprehended as we proceed. I will refer to four passages of the patent, to show that this is an exact statement of his theory. The first is on p. 2, folio 6. He says: "From this consideration, when taken alone, it would appear to be best, to place the axles as near to each other as possible, thus causing them to approach more nearly to the direction of the radii of the curves, and the planes of the wheels to conform to the line of the rails." The next passage is at the end of folio 13, and at the beginning of folio 14. He there says. "For this purpose, I construct two bearing carriages, each with four wheels, which are to sustain the body of the passenger, or other car, by placing one of them at or near each end of it, in a way to be presently described. The two wheels on either side of the carriages are to be placed very near to each other, the spaces between their flanges need be no greater than is necessary to prevent their contact with each other." I will now turn your attention to p. 5, folio 21, where he says: "The end which I have in view, may nevertheless be obtained by constructing the bearing carriage in any of the modes usually practised, provided, that the fore and hind wheels of each of them be placed near together, because the closeness of the fore and hind wheels of each bearing carriage, taken in connection with the use of the two bearing carriages, coupled remotely from each other as can conveniently be done, for the support of one body, with a view to the objects, and on the principles herein set forth, is considered by me as a most important feature of my invention."

You see that he says, "*the closeness of the fore and hind wheels*" is considered a most important feature of his invention. The next passage is in folio 23, where it reads. "The two wheels on either side of one of the bearing carriages may, from their proximity, be considered as acting like a single wheel." I will now read again the passage to which I first referred on p. 2. "From this consideration, when taken alone, it would appear to be best to place the axles as near to each other as possible, thus causing them to approach more nearly to the direction of the radii of the curves, and the planes of the wheels to conform to the line of the rails." There is the beginning and introduction of the philosophy of Winans, and the idea or principle is repeated four times in different parts of the patent. And I have read them to show that this is precisely the fundamental idea of all his arrangements. So far as regards this part of his structure, the predominant notion in his mind was, that the curves could be passed more easily, and a better car made, if the axles of the trucks were only brought as close together as you could possibly make them run. The whole patent is founded, and the whole invention is moulded upon exactly that notion. Let me ask your Honor's attention to the second branch of the theory; for it embraces two parts. The second branch relates to the *connection of the two trucks with the body*. Both these theories are the subject of the Plaintiff's claim.

I will refer to a few passages in the patent, which describe the peculiarity of Winans's mode of connecting the body with the trucks. I read from folio 16, of the Patent. "Upon this first bolster, I place another of equal strength, and connect the two

CONSTRUCTION. 39

together by a centre pin, or bolt, passing down through them, and thus allowing them to swivel or turn upon each other, in the manner of the front bolster of a common road wagon." I will also read the sentence preceding. "This bolster must be of sufficient strength to bear a load UPON ITS CENTRE of four or five tons." Your Honor will take notice that it says "*upon its centre*," for that is important. Then again in folio 23. "The bearing of the load on THE CENTRE of the bolster, which also is the centre of each bearing carriage, likewise affords great relief," (see how much importance is attributed to this, in carrying out Winans's idea) "from the shocks occasioned by the percussion of the wheels on protuberant parts of the rails, and other objects, and from the vibrations consequent to the use of coned wheels; as the lateral and vertical movements of the body of the car resulting from the above causes are much diminished." And there is another part of his patent where he states, that he wishes to have his bolster swivel, "in the manner of a common road wagon." Now everybody knows, that there is nothing new in the swivelling of a common road wagon, nor in suspending the weight of the load upon that part of the bolster which is close to the centre-pin. Nor is it asserted by the patentee that there is any thing new in it; and yet he adopted that, as an essential and elemental part in his theory of construction.

The next paragraph to which I refer is found in folio 20. " When the bolsters of the bearing carriages are placed under the extreme ends of the body, the relief from shocks and concussions, and from lateral vibrations, is greater than it is when the bolsters are placed between the middle and the ends of the body, and "— (this is the important passage)—"this relief is not materially varied by increasing or diminishing the length of body, while the extreme ends of it continue to rest on the bolsters of the bearing carriages, the load being supposed to be equally distributed over the entire length of the body." According to his idea or theory, the length or shortness of the body made no difference, provided it was suspended at the extreme ends. Hence the injustice of claiming as Winans's plan, or as any part of his *idea*, the use of a VERY LONG body; such as is now in common use, and such as his experts and his counsel *claim* as one of the important and distinctive features of his invention.

Such then is *Winans's theory* on the two branches of his claim.

I proceed now to say a word further in regard to the legal CONSTRUCTION of his claim.

Is not *this* the true construction of the claims of Winans? Namely, *The coupling of the wheels in each truck, so that the flanges shall come very close together, and the two trucks at or near the end of the body by* MEANS PECULIAR *to him*, viz: *by long springs bolted to the tops of the boxes of the axles, and bolsters bolted across the tops of the springs; and swivelling like a common road wagon, and bearing the load on the centre of the bolsters; dispensing with side bearings?* That is the structure of the car that Mr. Winans specially described and recommended as *the best* to embody his ideas. *Id est*: making "such an adjustment of the wheels and axles" (to use his own language) as will give a new ease of motion to the car, by the peculiar uses of the SPRING TRUCK, and connecting *the truck* with the body by the *wagon bolster*, dispensing with the side-bearings and the *thoroughbrace*, which side bearings, without extra springs, would give a hard motion to the car.

The *theoretical* idea was very GOOD on paper, but the *practical* trial was very *disastrous.* Yet is not this his claim? Can any man *deny*, that that, which he has specifically pointed out and recommended, is not his claim? Such was Winans's *theory*, such his OBJECTS, and such his peculiar mode of *embodying* that theory.

As this *matter of construction* is so *vital* in this case, and such extraordinary ideas have been entertained from the day when this litigation began before the Chief Justice of the United States, to the present hour, and so remarkable have been the shifting grounds upon which the patent has from time to time been put, that I propose to throw out some suggestions of argument upon the constructions that I have stated as the true construction of Mr. Winans's claim, and the only construction upon which the patent can stand consistently with any pretence that Winans was the first and original inventor of the things claimed by him. Suppose that the patent is *not confined* to the mechanical contrivances of LONG SPRINGS, bolted to the boxes of the axles, with *road wagon bolsters*, bolted across the middle of the springs, *as the means of arranging and connecting* the wheels with each other and with the body. And suppose that by the true construction of the patent, it is for *that mode* or *any other mode or mechanism*—arranged without frame or otherwise—by which the wheels could be connected closely in pairs to form bearing carriages, and the bearing carriages remotely from each other, to sustain the body. Then in that case the patent would be for the *principle* of the eight-wheel car, reduced to practice not only by the mode described, but by any other mode that will embody the same principle, *and therefore void*.

1st. Because the specific mode described and recommended is pernicious as a mode, and would require experiment to discover the error in the specification, and to develop other modes better adapted to embody that principle *not* described in the specification.

2d. It is void for want of originality, because the *true* principle of the eight-wheel car was prior to Winans's patent, described and *used* and embodied in mechanism different from Winans's, and the same as that now in general use.

3d. That if it could be deduced, *from the Plaintiff's specification*, that he *intended to place the wheels at a distance from each other, different from what was before known and used;* such difference would *not* be the discovery of a NEW PRINCIPLE, and it is not clearly definite and distinct from previously known distances.

And, as a difference in distance is a difference in *proportion*, and an incident common to all machines, such variation of distance is *not* a matter of invention, and does not develop any new physical law or principle.

It does not introduce any new elements, and form a new combination.

It does not constitute a new machine in its substantial parts, or in its essential character.

It is not the subject of a patent. You cannot patent a principle, and you cannot claim all modes of developing a principle.

Now, as the principle of the eight-wheel car existed in Chapman, Tredgold, the Wood car, the Trussell car; in Fairlamb's patent, in the Quincy car, in the Columbus, and in all the *old cars*, Winans's patent IS VOID, if he intended to cover *the principle* of any of them, because it is *too broad*. So said Judge Conkling.

We shall presently show, that whichever construction is taken by the Court, it is equally safe for the Defendants. For under the *first construction*, we do not *infringe;* and under the second, the PATENT is VOID.

If the Plaintiff could sustain a patent for ANY MECHANISM, coupling the two wheels close together, and placing the trucks at the ends of the body, notwithstanding the prior existence of the Chapman, Tredgold, Wood, and Trussell cars, the Columbus, the Winchester, &c., and the Allen engine, then the Defendants do *not infringe*, because THEY place their wheels FURTHER APART than the Wood and Trussell cars, and the Columbus and the Winchester, &c.; and *substantially* AS far apart as Chapman, Tredgold, Quincy and Allen; and *much further apart* than the Plaintiff allows them to be placed.

Suppose the Plaintiff means to claim as his invention " the before-described manner of arranging and connecting the eight wheels so as to accomplish the end proposed, that is—to run smoothly—*by the means set forth*, OR ANY OTHERS WHICH ARE ANALOGOUS and dependent on the same principles; this patent IS VOID, because the Plaintiff claims TOO MUCH. If change of *proportions* prevents the preceding eight-wheel from being *analogous* or dependent on the same principles as Winans's, we do NOT INFRINGE.

The plaintiff must STRIKE through Chapman, Tredgold, Quincy, Allen, Wood and the Trussell cars, because Winans's is at the *minimum distance*, and the Defendant at the *maximum distance;* and the *prior* inventions are between them both. The first makes the Plaintiff's patent VOID, or else the Defendants are not infringers. In other words: Winans embodies the theory of conforming the axles to the mean radial line of the curves, while the preceding inventors *did not*, and the Defendants DO NOT.

Although Winans's theory be ever so distinct and clear, yet, if the Plaintiff could claim a Patent for *nearness* of wheels in *one truck*, it would be equal to claiming a *patent* for putting a

CONSTRUCTION. 43

large wheel in place of a *small* wheel—for your Honor will see by the models, that by putting larger wheels on the same trucks, you bring them closer together.

Again. If the Plaintiff could claim *remoteness* of *two trucks,* *lengthening* of the *body* is all that is wanted to produce that remoteness. It is a mere *change of proportion* alone, and *nothing else.* With these remarks, upon what is not—I pass again to WHAT IS the *legal construction* to be given to Mr. Winans's claim.

Taking *into view* the fact, that there were *other eight-wheel double truck* Railroad cars for *freight,* if not for passengers, existing many years before his patent, and capable of carrying passengers as well as freight; *also,* that the patent must, by law, be so construed, if possible, " *ut magis valeat quam pereat* "—that if the construction *can,* without violence to the language, it should be so made as not to render the patent VOID, as embracing that which Winans did not invent. *Also,* considering, that *while* the placing of the wheels near together in each truck, and the trucks far apart, was a mere change of proportions, and was not original with Winans, and he can have no exclusive right to use that mode of arrangement, yet he was *the first* to *couple* the *axles* with a *long spring,* and to use the wagon bolster, which nobody ever used *before* or *since* the date of his patent, Winans's claim, if he has any *valid* one, must be *confined* to *these peculiarities of mechanical structure,* leaving *others* to couple their wheels as they please ; and to use the rigid rectangular wheel-frames, with one spring to each wheel, as was done of *old.* And the " *analogous modes,* dependent on the *same principles,*" must be confined to any other mode of *connecting the axles* by some *elastic* material, so as to allow of the wheels separating and approaching each other as the body rises up and down.

And, as Winans *cannot* patent his THEORY, he can claim the peculiar *devices* by which he embodied that theory,—springs, &c., if they were first invented by him, as distinguished from what was known and *used* before.

Now the *other* theory, as to the *connection* of the trucks with the body, was to carry the body more *smoothly,* by *suspending* it in a *new way,* so as to give it *new freedom* of motion upon *springs.* 1st. By placing the bolsters on *long* elastic springs. 2d. By resting the *whole weight* of the body upon the *central part of the bolster,* (and not on side bearings,) the manner of a common road wagon.

This theory also required the *peculiar devices* above stated. See patent, p. 6, where it says; "The bearing of the load on the centre of the bolster, which also is the centre of each bearing carriage, likewise affords great relief from the shocks occasioned by the percussion of the wheels on protuberant parts of the rails or other objects, and from the vibrations consequent to the use of coned wheels."

Now let us look at the theory of the Plaintiff's counsel, as to construction.

They admit and assert, that the claim is not for any part of the structure, unless combined with some apparatus for drawing the car by the body. If I am right, there is an end of their claim.

Mr. Keller. Permit me. We say that the drawing by the body was the only mode by which the principle of the invention could be practically carried out.

Mr. Whiting. That is what I mean precisely. It is a vital principle; and if you do not draw by the body, you do not use that which embodies the vital principle of the Plaintiff. They also say, that it *is not for the near coupling* of the *wheels*, and the distant position of the two trucks and their connection with the body, for the *purposes stated*, UNLESS *combined with some apparatus for drawing the car by the body*.

This is the substance of their claim they say, and if their patent does not embrace that claim, they have no claim that will stand.

They do not claim the drawing by the body as the Plaintiff's invention; but their claim is to be deemed applicable only to cars which are drawn by the body, and *not* by the perch.

So that, if the patent covers cars which draw by the perch, such a construction of the car is worthless, and does not embody the principle of the Plaintiff's invention, and the use of such a car is not an infringement of the real claims of the patent, as the Plaintiff's counsel understand them. And in such a case, the patent would be void, because it would cover prior inventions.

Now, my conclusions are these:—

1st. That this admission is equivalent to an abandonment of all claim to novelty of every part, and every combination of parts of the eight-wheel car, except when *used in this way*.

2d. As this mode of drawing is not mentioned, or described, or claimed in the patent, it results from the Plaintiff's admission that we may use every thing that the patentee has described or claimed, *without* infringement of his invention ; and we become infringers only when we use something *not* there described or alluded to in the remotest manner! This is the most remarkable doctrine that I have ever listened to in a patent case. And an attempt, however adroit, to smuggle into the patent that theory of construction, and those ingenious claims which are not found in the language of the specification, nor in the intentions of the patentee, and which are palpably inconsistent with both, cannot be successful.

The learned counsel are driven to the use of vague phraseology, because it is impossible, in clear language and upon settled principles of construction, to discriminate between what *Winans*, —(using the language of the specification,)—really claims as distinguishing his invention, from what was well known before, without, at the same time, so narrowing his claim as to render him unable to make out a case of infringement.

Hence, instead of using terms which are familiar to patent lawyers, instead of claiming either the specific parts, or *some combination* of the parts of the machine for transporting freight or passengers, he says his claim is *not* for the parts, nor for any combination or arrangement of parts, but for *a whole!*

A patent for an improvement on any machine, be it a car or not, must be either for some new part, or some *combination of parts*, whether new or old; and that combination may embrace all the parts or only a certain number of parts. But when, as in this case, all the parts were old, most, if not all, the combinations of these parts were old, the Plaintiff can claim only such *combination* as is *new*. He can claim nothing more or less than *some* combination. He cannot claim the CAR AS A WHOLE. Otherwise, he would cover many parts and many combinations which it is admitted he did not invent, and so his patent would be void.

The true construction is as I have before stated, and in accordance with the opinions of the learned Judges who have had the claim under judicial consideration. See Chief Justice Taney's construction, Plaintiff's book, p. 20, (at the end;) and which has been adopted by Judge Conkling and your Honor; in which he says,—" But he claims as his invention, the MANNER *of arranging* and *connecting the eight wheels*, as specified in his patent, for the end above mentioned. And also the *connection* of a railroad *carriage body with them*, adapted either to the transportation of merchandise or of passengers."

The object of the invention has been already stated; and the *means* by which it is to be attained are classified thus in the language of the patentee. 1st. *By the beforementioned desideratum of combining the advantages of the near and distant coupling of the axles; and* 2d. *By other means to be hereinafter described.*

The ONLY MEANS stated or described in any part of the patent, are as follows :—

1st. By constructing the spring bearing-carriage with the wheels very close together.

2d. By the distant coupling of the trucks, so that the body is supported at or near the ends.

3d. That the body be twice the length of the ordinary four-wheel car; but the length is stated to be immaterial.

4th. The trucks swivelling in the manner of a common road wagon.

5th. The bearing of the load on the *centre* of the bolster.

There is no part of the patent where any thing is said about the means of draft. There is nothing said or intimated as to the absolute freedom of swivelling as being necessary or desirable. The subject itself is not alluded to in any part of the patent.

On the contrary. Two facts are mentioned expressly, which are inconsistent with the pretension, that the great idea of the patent is absolute freedom of swivelling.

1st. The trucks are to be " *coupled*," as well as the axles of each bearing carriage or truck.

Then, 2d. The two trucks are to " *swivel*," in the *manner* of a " *common road wagon*." They are not to be left to fly round " *ad libitum*." Each truck was to act *substantiallg like* a single pair of wheels; which every one knows could not run on a road without the guidance of a perch.

3d. The patentee says, that all the advantages contemplated by him may be obtained by substituting, instead of his spring-truck, any of the ordinary bearing carriages; and those *all* drew *by* the *perch*.

There is not only absolutely nothing in the patent to sustain the idea put forth by Mr. Keller, that absolute freedom of swivelling was essential to the construction of the car as described in the patent, but the whole patent is directly in the teeth of it.

Therefore the Plaintiff's patent cannot be construed, as embracing or embodying any idea of the absolute free swivelling of the trucks; nor as applicable only to such an arrangement; nor as being distinguished from other inventions by any such criterion.

Thus far we have shown what we believe to be the legal construction of Winans's claims.

1st. By considering the admissions of the patentee, in his specification, and an analysis thereof.

2d. By a close examination of the language of the specification and claim.

3d. By judicial authority.

Now, there is left only, on the subject of construction, and to throw light upon it, a consideration of the state of the art of making the running gear of the eight-wheel cars, at the time of the alleged invention.

This is important, both to the Patentee and to the Defendants. But the statement of the evidence must be postponed till we come to the proposition, "That Winans was *not* the first inventor of the first '*double truck eight-wheel car*,'" such as is used by the Defendants.

DISCRIMINATION.

The next point to which I ask your Honor's attention is this. That the *patent is not valid, except for the peculiar devices described in the patent, because, being an improvement on what was known and in use before, the patentee is bound to discriminate* between his own improvements, and the cars nearest like his own, viz.: *the preceding eight-wheel cars.*

The questions arising upon this subject at the present trial are very different from those which have been formerly discussed. On previous occasions, the only questions which arose were simply whether the Plaintiff had, in his specification and claim, sufficiently discriminated between the eight-wheel car, described by him, and the old *four*-wheel cars, and the Courts held that the Plaintiff had done so. Now we have introduced evidence of other *eight*-wheel cars prior to the Plaintiff's, which embodied the main features of that which the Plaintiff describes. And the issue now is an entirely new one, arising upon a new state of facts. I should not have argued this question before your Honor, if this were not so. The new evidence to which I allude is, that there were sundry eight-wheel double truck swivelling cars in existence prior to the Plaintiff's patent, which had not been proved on any former trial.

Now, if the description is not sufficient to discriminate his invention from others, that description is too vague and ambiguous to come within the requirements of the patent law.

Next. If his claims do not also clearly distinguish between new and old, they are void; because the patent is too broad.

It is a *fact* that many eight-wheel cars were in use on the Baltimore and Ohio Railroad, from the fall of 1830 to October 1st, 1834, inclusive, not invented by Winans. To say nothing of the Columbus, the Wood and Trussell cars, there were the Dromedary, the Winchester, and other passenger cars; also the Quincy cars, and the Allen engine, &c.

In what particulars was Winans bound to show the *difference* between what he claims and those that we show were in use, and upon which his improvements were founded?

We say that he was bound to *discriminate* between what was *old* and what he claimed as *new*, in all the particulars in which he claimed that *his invention consisted.* What he does *not* thus discriminate is admitted to be old, or not his invention.

And what has he discriminated and recommended? It is only the *spring truck* and the *wagon bolsters.*

In this alone has he conformed to the requirements of law; and for this *alone*, therefore, is his patent valid. What difficulty was there in Winans being definite, and in discriminating between his pretended claim and others? If his body were twice the length of the ordinary four-wheel car;—(i. e. from 20 to 24 feet,)—why not tell how long other eight-wheel cars, for passengers and freight, as the Quincy and Trussell cars, &c., *had been?*

If he placed the bolsters differently from others, why not say so; and tell how far from the end of the body they must be placed in order to come within his plan?

As to the distance of the wheels in the truck, others had placed them 3½ feet apart, as he *admits*. Is that too far? If so, why not say so? In truth, others had already placed them as near as Winans did, and, therefore, he dared not specify the distance, although his counsel now claim that he did in fact introduce a substantive improvement, by diminishing the distance between the flanges.

Why not show what he meant to claim by some original drawing? None has been proved, or referred to, and the inference that we draw is, that none really existed.

The present drawings do not help the difficulty. They were not made till 1838, as preparatory to the suit of Winans vs. The Newcastle and Frenchtown Railroad Company.

The drawing before the Court, representing a car that was not built in 1837, does not in any way illustrate the specification, but embraces things *not* in the patent, and portrays modes of construction entirely the reverse of what is there described.

I shall ask your Honor's attention to numerous particulars, in which there are diversities between the drawing and the patent. And your Honor will have to settle first the question;—What will be the legal effect of a drawing, no part of the original patent, and not restored when the patent was restored, and not made until it became necessary for the purpose of founding a lawsuit; made, therefore, "*pendente lite,*" and portraying a car that was not in existence until four years after the date of the patent, whereby it is attempted to grasp and comprehend things which are not in the patent; and not that only, but also to show things which are directly inconsistent with what is contained in the specification, and directly the reverse of what the patent prescribes?

(Presenting the drawing to the Court.)

This drawing represents a freight car and not a passenger car, which is the only one described in Mr. Winans's specification. The drawing represents a rigid rectangular wheel frame for the trucks. In the specification, instead of a rigid wheel frame, the axles of the wheels were connected only by a steel spring, bolted to the boxes of the axles, with a bolster bolted across to the tops of the springs. The drawing represents two springs on each side of the truck frame, the action and reaction of which may not tend to throw the axes of the wheels out of parallelism. The specification particularly recommends one spring only on each side of the truck to connect the axes of the wheels, and the action of that spring would necessarily throw the axes out of parallelism. The drawing represents the springs with the shorter leaves downwards. . The specification directs exactly the reverse.

In the drawing, the bolsters on which the body rests are placed between five and six feet from the ends of the platform of the car; whereas the specification requires the same to be placed at or near or beyond the ends of the body, and in any event no farther under the body than that the wheels shall come just within the ends, and the trucks are to be coupled as remotely from each other as can conveniently be done for the support of one body. In the drawing the wheels are placed sufficiently far apart to put a brake in between them; while in the specification the wheels are directed to be as close as possible, without the flanges touching, to have them act as near as may be like a single wheel.

The drawings represent a conical pivot marked X, with sockets and side bearings, forming a solid bolster in one solid piece, with a lower bolster and pocket Y to correspond—while the specification describes a plain bolster of wood or iron, reaching across from spring to spring, united to an upper bolster by a king bolt swivelling in the manner of the front bolster of a common road wagon.

The drawing shows a mode of coupling or drawing the car by two pieces bolted across the bottom framing, and a coupling bolt with a ring to it to drop through the coupling, to draw the car from the middle of the end of the body—the specification neither describes nor intimates any mode whatever by which the cars are to be drawn. The drawing shows cast iron pockets for the ends of the springs to work in—the specification prescribes a different mode of fastening the ends of the springs, viz. :—bolting the ends of them on to the boxes of the axles. The drawing shows an arrangement of brakes suited to the swivelling trucks of the eight-wheel car—the specification does not describe nor mention any mode of arranging or using brakes.

Now, what are we to do with a drawing that comes so directly across the track of the patent?

The patent is not to be enlarged by reference to the drawing.

The drawing does not help the Plaintiff upon these defects of discrimination. The specification must always govern the patent.

There is nothing in the drawing which shows how other preceding cars were made, so as to aid the specification in its fatal defect of non discrimination. If the Plaintiff meant to cover any thing beyond the elastic spring trucks, why should he not have said so?

Why not discriminate between his invention and the other preceding eight-wheel cars, viz. :—

1. The Chapman.
2. The Tredgold.
3. The Allen Carriage.
4. The Quincy Cars.
5. The Wood Cars.
6. The Trussel Cars, for horses and carriages and soldiers.
7. The Columbus, the Dromedary, &c., and also the Victory, at Philadelphia?

These all had running gear adapted to passenger cars, as well as for freight, and were the same as those now used; (and Winans claims *freight* as well as passenger cars.)

We say that there was no patentable discrimination to be made, excepting as to the *peculiar devices* by which Winans embodied his theory, which were :—

1st. The approximate coincidence of the axles with the radii of the curves.

2d. The central bearing of the bolsters.

And finally as to discrimination. If the Columbus were invented by Winans, and the Winchester and Dromedary and other eight-wheel cars, being allowed to go into public use *before* the patent, with Winans's knowledge and *consent*, they became public property (the statute of the 3d of March, 1839, not applying to this case).

Being public property, if they did embody Winans's pretended invention, then the Plaintiff's patent is void, *by reason of abandonment*. If they did *not* embody Winans's invention, the Plaintiff was bound to discriminate between them and his own invention, or else his patent is void for want of discrimination.

His patent does *not* discriminate except in respect to *spring trucks* and *wagon bolsters ; these* are, therefore, all he can claim; and *these* the Defendants do *not* use. Your Honor sees the Defendants are safe in either alternative.

ABANDONMENT.

I now pass to another subject, on which the evidence is somewhat voluminous, but which it will not be necessary to recite to the Court.

It is this : *That if such running gear as the Defendants build, is embraced within the true meaning of the Plaintiff's patent; and if Winans were, in fact, the first and original inventor thereof, yet his letters patent are void in law, said invention having gone into public use, with his knowledge, and without objection on his part, prior to the date of his patent.*

If a man stand by, and see his own property sold by a person having no title, he loses all legal right to reclaim it from the vendee. So a patentee, allowing his invention to go into public use without objection, virtually admits the right of the public to use it. This is just. Otherwise, he might be permitted to lie by, until many innocent parties had rendered themselves liable as infringers, had invested their capital in the business, and thus put themselves in the power of the patentee, and be liable either to utter ruin, or to most exorbitant *damages.*

The act of Congress, in force before 1834, provided, that a patentee, who had allowed his invention to go into public use *at any time before* the date of his patent, thereby lost his exclusive right to his invention ; and therefore the patent was void. The statute of 1839 provides, that public use of the thing patented must be for two years preceding the date of the application for a *patent*, in order to be construed as an abandonment. This last statute is not applicable to the present case. See Chief Justice Taney's opinion. The reason of the rule is, that dedication *once* made, *cannot* be *revoked*.

Now, what are the facts ? None dispute that the Columbus, the Winchester, the Comet, and three or four other eight-wheel double truck railroad passenger cars were in open and public use upon the Baltimore and Ohio Railroad, both previously to and at the time when this patent was taken out; which was in 1834. I will read the extract from the Report of Mr. Gillingham, who was the superintendent of machinery on the Baltimore and Ohio Railroad, in the eighth Annual Report of the President and Directors, to the Stockholders of that Road, made October 1st, 1834.

At page 32, vol. 2, he says:

"Four new passenger cars have been constructed during the present year, viz.

"1st. The 'Winchester,' carrying thirty-six passengers, on eight wheels.

"2d. The 'Dromedary,' a large and commodious car, on eight wheels.

"3d. The 'Comet,' a car with five bodies, carrying forty passengers, on eight wheels.

4th. The 'Patterson,' on four wheels.

"*Four of the old cars have been repaired, and placed upon eight wheels.*"

There were then these eight eight-wheel cars in use on that road at this time, which was before October, 1834.

Now, if Winans invented the "Columbus," the "Winchester," the "Dromedary," and "Comet," they being eight-wheel cars, then these were allowed to go into public use, with his consent, and by contract; and also, *necessarily*, with his knowledge and consent, the great number of Wood cars, and Trussel cars, that were built after the Columbus, together with those built before the Columbus, also these other four-wheel cars, were never pretended by any witness to have been experimental cars.

If these cars had their wheels near together and coupled far apart, the trucks being the common bearing carriages, as referred to in the patent, they embodied Winans's pretended invention substantially; and so completely, that he cannot discriminate between them and his.

Therefore Winans's pretended invention was dedicated to the public, not merely for a few months, but for nearly four years, as we have already shown.

Unless Winans clearly discriminates between what he claims, (by a fair construction of the terms of the specification,) and these preceding running gears, or, in other words, if his patent is sufficient to cover running gear such as they had, then the invention described in the patent was in public use, and the abandonment complete.

But the Plaintiff says, that although these cars were used on a Railroad, yet that use was for the purpose of EXPERIMENT only, and that such use, although continued for so long a time, is not to be construed as an abandonment of the invention. Then the question is, whether the use of so many eight-wheel cars for so long time, in such a way, can properly be considered as a use for the purposes of experiment *only*, or is it a use not merely for the purpose of experiment, but for the ordinary business of the Railroad on which they were placed!

What does it mean to experiment? The word comes from the good old Latin " *Experior*"—*to try and find out something*, and something not settled and familiar. And it is vain for the Plaintiff to say, that he was experimenting four years; because far too many cars, both for passengers and for freight, were in public use, to be considered as an experiment.

What did Winans do, that looked like experimenting?

If the Plaintiff pretends that he was experimenting, to discover a mode of drawing the car, this cannot be true; for the specification does not describe any mode of drawing as known or discovered, or as having been the subject of experiment.

The mode of drawing now used, and shewn on the restored drawing, was first suggested in 1835, by Jacob Rupp, after the specification was drawn and the patent issued.

The last car built and tried, previously to or during the making of the specification, was the "Comet," which drew by the perch, and which car the patent describes, without suggesting any mode of drawing. Moreover, experiment was not necessary to discover the mode of drawing by the body, because that mode was described in all the books long before 1830.

How does it happen that so much experiment was necessary in the enlightened city of Baltimore, to find out how to make a good car, when Dorsey could build one at once, he having seen nothing else but the alleged *failed* experiments at Baltimore?

He was at once successful, without experiment.

Their experiments marched like the apparent course of some of the comets, more with a retrograde than a progressive motion.

The "Victory" also, in Philadelphia, was successful at once, without experiment. The invention leapt from the head of its author all perfect and complete.

How many years was it necessary that the Columbus should continue to run off the track, to complete the experiment?

Still the Plaintiff insists, that the car Columbus was an experiment merely, used from July 4, 1831 to October 1, 1834—three years and a quarter—and other cars followed in 1833 and 4, two or three years after, and he asserts that he had a right, while trying these experiments, to put these cars in public use, without losing his right to a patent.

My answer to that is, that it is *not* an experiment merely because the *use* was *continued* for several years. That seven other passenger cars followed in gradual succession, not for any such purpose as to try experiments, but as the business of the road called for them. If such use of eight different cars can be made, without being held to be a *public* use, any inventor may put his machines into use, and run them year after year, and pretend it is an experiment, provided that he at last takes out a patent. That is not the true understanding and construction of the Patent Law. I admit, that a man shall have a reasonable time to try his invention; and if it be required, that he shall use it before the eyes of all mankind while perfecting it. And if a man have a patent for a wagon, he may satisfy himself by making a reasonable number of experiments with it upon the public roads; and the fact of his driving on a public road is immaterial; and if it be an experiment on a Railroad car, that it shall run upon a Railroad, for that will not constitute the abandonment of his invention. But, Sir, there is reason in all things. A man has no business to let loose upon the public his crude and incomplete notions and half-finished inventions—to permit them to go into public use, and thus become public property, and by-and-by undertake to recall that which is no longer his own, because some lucky idea has given a new value to an old invention, which was not, in its original form, worth the expense of a patent.

But the same description of trucks, which the Plaintiff claims, (excepting only the PECULIARITY OF SPRINGS) were used under a variety of cars long before the Plaintiff's invention; and the *claim* in the patent relates only to the trucks, or running gear. The *only experiments*, therefore, which can have any connection with *this patent*, must be those which related to the *running gear or trucks*, and their connection with the body.

Now there is *no evidence* of any series of experiments having been tried on this *running gear*, whereby the *results* attained and described in the patent, vary in any *substantial* particular from the result attained in the Columbus. So that whatever experiments may have been tried as to the former size, shape or length of the body; or even as to the position of the trucks, no experiments can be shewn to have been made with *special reference* to that part of the car which is the subject of the Plaintiff's claims. And therefore experiments on other parts, or for other purposes, could not be such experiments as would be an answer to our claim of abandonment.

We must not forget what is the subject of our investigation. We must not use the word " *car* " in the sense that the patentees' witnesses have given to it. The "car" means the body and the trucks, and I know that they have experimented a long time on the construction of the body; but, in regard to the running gear, they ceased to perform their experiments after the Columbus had run a short time.

Your Honor will recollect an important piece of testimony on this subject, and that is, the contract which Winans made, and on which an opinion was given by the Chief Justice.

The testimony of Mr. Thomas also shows, that the contract was in force from the time that the Columbus was put in use, until the date of the patent, and also that there was a verbal agreement besides the written one, that *all* the inventions made by Winans should belong to the company, or be subject to their use. Now the Comet was the last car built before the issuing of the patent, and *not* the Washington cars. These were *not* built, and no experiment was tried on them before the patent was issued. Now, Sir, was the running gear on the Comet an improvement on that which had long been in public use on that road? Was it an embodiment of some new principle? No. On the contrary, so far as relates to the running parts and their connection with the body, they were precisely the same as those which had previously been in public use upon a variety of Wood and Trussell cars, and other eight-wheel cars. I am aware that it has been alleged that there were changes made as to the nearness of the wheels, but these changes, taken alone, and considered by themselves, are not deemed even by the learned counsel of the Complainant as material. And I challenge them to point out a *change in the running gear*, which they dare to say to your Honor is a change in its mechanical principles, or mode of operation, or one which is *material* and ESSENTIAL, and not formal.

If Winans made experiments *three years*, and tried to improve upon former eight-wheel double truck cars, why did he not in his patent, inform the public clearly, what was the result of his experiments?

The language of the patent covers the Columbus, and all the other cars, as well as the Wood and Trussell and Quincy cars, etc., then why did he claim that which he had found to be unsuccessful?

Experiment of three and a quarter years duration, did not develop any thing as to the running part of the Columbus. How easy would it have been to alter the running part, if any alteration was necessary, while they did alter other parts of the car. If the arrangement of the running part of the Columbus did not embody the principle which Winans claims, he was bound to discriminate between that and what he did claim; otherwise the public are left in the dark. And not having discriminated, but having finally claimed the precise arrangement used in the Columbus, after the same arrangement has been used for years, on the Wood and Trussell cars, etc., and on the Columbus itself, this fact shows conclusively that the arrangement of the wheels, and the connection of the trucks with the body, which were the only subjects of the claim, were not a matter of experiment, but were left in October, 1834, just where they were in July 4, 1831.

In the cars of the Defendants, the wheels in each truck are more distant, and the trucks farther from the ends of the bodies than in the Columbus, and the body far more resembles the Columbus than the Dromedary or the Comet, so that if the patent covers the cars of the Defendants, it must also cover the Columbus; and if the Columbus did not embody the Plaintiff's principle, then the Defendants' cars do not.

If the Plaintiff contends that his experiment was to ascertain the proper position of the truck under the body, and that the Columbus was wrong because they were placed too far from the ends of the body, we answer:

1st. That the statement is not true.

2d. That the position of the trucks in the Defendants' cars is farther from the end of the body than in the Columbus.

3d. That all the different positions of the trucks, both at and near the ends of the body, had been used before the Columbus on the Wood cars, on the Trussell cars, and on the Quincy cars. All points of support at or near the end had been used, and that embodied in the Columbus is nearer like that which is now in use than Winans's.

4th. This distance of trucks must vary according to the proportions of the body. (See Winans's contract with the Reading Railroad Company, which shows an instance where the eight-wheels must be brought close together, in order to place two swivelling trucks under one locomotive.) Allen did the same thing.

The fact of allowing these running parts to go into use on the Baltimore and Ohio Railroad, with Winans's knowledge and consent, it being under contract, is *conclusive evidence of abandonment.* I will refer your Honor to the opinion of Chief Justice Taney on this subject, and add that it is proved in this case clearly, and admitted, (what indeed was not admitted before) that there *were* double-truck swivelling timber cars in use before the Plaintiff's patent, and that they were long-bodied, if you choose so to call them. Your Honor will not forget the elaborate, the philosophical, and the admirable exposition of the principles of physics involved in that structure, which was given the day before yesterday by my learned colleague.* Applying the instructions of the Chief Justice to that subject, it seems to me that the Court must decide on this point in favor of the Defendants, unless the judgment of the Chief Justice is erroneous; and your Honor cannot fail to perceive how much stronger the evidence of *public use* of the running gear now is, than it was when that case was tried.

If the gentlemen undertake to tell us that there was something very peculiar in the structure of the Baltimore and Ohio Railroad, I answer: that the curves were no sharper than on other roads. The curves of the road on which the Quincy cars ran were the sharpest, not exceeding in some instances 150 feet radius. And on the Baltimore and Ohio Railroad there are none of less than 400 feet radius. There is no curve now used on any roads so sharp as those formed by the switches, and no better arrangement of wheels was required than that contained in the Wood and Trussell cars after they were put into successful operation.

The increase of speed required that the wheels of each truck should be put farther apart, not that they should be brought closer together.

Now Sir, what seems to us as quite conclusive on this subject of experiment is, that no such car, as Mr. Ross Winans has claimed to be peculiarly his, has ever been put into successful operation on any road in the country. On the contrary, science and experiment have shown the fallacy of his whole theory, and the uselessness of his construction. All experiment has established this fact, that what I shall call the " *Simon Pure*" Winans's Car, has undergone but one experiment, and that ended in its abandonment and destruction. I refer to the testimony of

---

* William W. Hubbell, Esq. of Philadelphia.

Schryack, *Worden* and *Shultz*; and these three witnesses, uncontradicted, have stated that to be the result of all the experiments on this car of Mr. Winans.

The Columbus was altered in 1832, then having new trucks placed under her—or, as the Plaintiff's witnesses say, having the wheels brought nearer in each truck, and the trucks farther apart—she then in 1832, embodied the principles of Winans, and being *publicly used* from that date to the date of the patent, there was an abandonment of the invention. The witnesses who prove these alterations are: " Michæl Glenn, Henry Shultz, John W. Eichelberger, Edward Gillingham, Conduce Gatch and Jacob Rupp; and almost all the Plaintiff's witnesses agree to the fact that the alteration was made, but they do not fix the date at which it was done. Indeed it is useless to talk about Winans's experiments so far as concerns the running part, since, in truth, long before the Columbus, and long before Winans pretends to have begun his experiments, the Wood and Trussell cars, in 1830, had embodied all the principles of the eight-wheel-car, in a form far nearer what the Defendants use, than are described by the Plaintiff.

Finally. Whether Winans was, or was not, experimenting for four years before he took out his patent, it is not *material, if* he allowed the Columbus to be in public use by a Railroad Company.

I believe that I may refer your Honor to the opinion of this Court on the motion for a new trial in the Troy case. See pp. 7, 8. The judgment, as to the question of abandonment, was founded upon the state of facts *then* presented in the evidence. And there the Court say, that " there were repeated failures in the experiments tried, and the cars abandoned, before the perfection of the car described in the patent." That was the statement pressed upon Judge Conkling's mind, and it was entirely a mistake. What evidence is there that any one of *these* cars was *abandoned*, or that *these* experiments resulted in failure?

On the contrary, it is proved that the cars were used in the ordinary business of the road,—no one knows how long; but I suppose until they were out of fashion.

These things show that the Winchester, the Dromedary and the Comet were in public use on the 1st of October, 1834, and not abandoned, and how the Court happened to be misinformed, I know not, in regard to the true state of facts on the subject of abandonment.

It was not the cars, but it was the alleged invention that was abandoned. And while these facts took place upon the Baltimore and Ohio Railroad, the use of the same double truck running gear had spread over the whole country, almost from Massachusetts to South Carolina.

The witnesses on whom we rely to prove the abandonment *before* the patent, by showing public use, with Winans's knowledge and without objection, are,

1. Philip E. Thomas. Troy case, p. 4. Answer to Int. 2.
2. Edward May.
3. Leonard Forest.
4. Conduce Gatch.
5. John Rupp.
6. Jacob Rupp.
7. John M. Eichelbergher.
8. Edward Gillingham.
9. Henry Shultz.
10. W. E. Rutter.
11. John P. Mittan.
12. The " Baltimore Patriot."
13. Winans's contract with the Baltimore and Ohio Railroad Company.

I will also refer to Chief Justice Taney's charge, and also the charge of Judge Conkling, in Defendants' No. 1, p. 46, fol. 304, on the subject of *acquiescence.*

I refer to this opinion because a question there arose, whether abandonment depended upon the *intention of the patentee*, and which may be important here.

To show that the question of abandonment *does not depend upon the intention* of the inventor, but that it does depend upon his acts or conduct, I refer to Curtis on Patents, p. 328—the note; also to the case of Shaw vs. Cooper, in 6th Peters, p. 218; and to the case in 7th Peters, pp. 292, 321—3; and to the opinion of his Honor, Judge Nelson, in McCormick's case, p. 12 of the pamphlet report.

I think that the Court will be satisfied, upon examining those authorities, that this is the doctrine of the Courts of the United States at the present moment. I, therefore, have nothing further to add upon the subject of abandonment of the alleged invention *before* the patent was taken out.

LACHES.

There is another fact of importance, viz. : That *after the patent was issued, it laid dormant, and the public were permitted to suppose and believe that Winans did not pretend to claim the ordinary eight-wheel car.* Its use became universal, and with Winans's knowledge, and with scarcely any objection.

The Defendants' witnesses on Winans's knowledge of eight-wheel cars being in use without claim or pretence of invention after the date of the patent, are :—

1.	Laban B. Proctor,	Defendants'	No. 3,	p.	8.
2.	John Murphy,	"	" 3,	"	9.
3.	Charles Minot,	"	" 3,	"	159.
4.	Wm. Raymond Lee,	"	" 1,	"	41.
5.	George Law,	"	" 2,	"	29.
6.	David Beggs,	"	" 2,	"	31.
7.	David Matthew,	"	" 2,	"	32.
8.	Albert Bridges,	"	" 1,	"	37.
	" "	"	" 3,	"	134.
9.	Timothy L. Smith,	"	" 1,	"	37.
10.	James L. Morris,	"	" 1,	"	39.
11.	Daniel N. Pickering,	"	" 1,	"	39.
12.	Charles Davenport,	"	" 1,		39, 40.
13.	George S. Griggs,	"	" 1,	"	40.
14.	Robert Higham,	"	" 1,	"	40.
	" "	"	" 3,	"	132.

Winans himself says, that *he* knew they were in general use in his bill.

15.	John Stephenson,	Defendants'	No. 1,	p.	42–3.
16.	Leonard R. Sargent,	"	" 1,	"	43–4.

17.	Jeremiah Van Rensselaer,	Defend.	No.	1, p.	44.
	" " "	"	"	3, "	136.
18.	Edward Martin,	"	"	1, "	45.
19.	Isaac Adams,	"	"	3, "	138.
20.	George Beach,	"	"	2, "	36.
21.	John Wilkinson,	"	"	2, "	43.

Under these circumstances, having thus allowed the running gear (which he now claims) to go into public use before the patent, and having made such slight claims up to 1847,—nearly the whole life-time of the patent,—stronger evidence of dedication or abandonment before, and of laches after the patent, could not be offered to any Court.

The next proposition on which the defence rests, I shall state briefly, but not argue. It is, *That the patent is void, because Winans's changes from what was known before, are bad in theory and pernicious, not useful in practice.* And I do not wish your Honor to understand that I lay great stress upon this point, because an invention may be, upon the whole, pernicious, and yet, if it were not intended to be pernicious, it may have SOME utility, and that utility may be sufficient to prevent the Court from pronouncing the patent void in law. Still, I have stated the point strongly, that the Court may see the bearing of the evidence that I intend to adduce, and which has a more important bearing on another part of the case. Not to occupy your Honor's time, I will furnish a list of witnesses who prove that Winans's theory in his specification is pernicious. This will dispose of that part of my subject.

Therefore, in recapitulation on the subject of the construction of the claims of the patent, we say:—

1st. That the patent is not valid, because the patentee has not discriminated between his own alleged invention and what was well known and in public use before; and that whether Winans did, or did *not* invent the Columbus, it is immaterial on this point.

2d. That the alleged invention,—if it be construed to embrace such cars as the Defendants use,—was in common and public use with the consent of the patentee *before*, as well as *after*, the date of his patent, and, therefore, he had no right to any patent, whether he was or was *not* the inventor; and upon this point, it is not material whether the Plaintiff was or was not the inventor of the Columbus.

3d. That the patent is void, because, so far as it introduces any changes from what was well known before, these changes were not useful in any degree, but pernicious.

NOVELTY.

The next main ground of defence is, that Ross Winans was not the first and original inventor of *that arrangement of the eight wheels, nor of that connection thereof with the body of the car*, which are used in the railroad cars built by Messrs. Eaton, Gilbert & Company.

To maintain these propositions, we shall rely:—

1st. Upon the conduct of the patentee, before and after the date of his patent.

2d. Upon the clear proof of numerous inventions of eight-wheel double truck railroad carriages, prior to the invention of the Columbus; also of many more, prior to October, 1834. We shall develop their history, their peculiarities of structure, and answer the criticisms made upon them by the Plaintiff's witnesses.

3d. That Ross Winans was *not* the inventor of the Columbus; but has attempted to impose upon the Court a fictitious, smoked drawing, in relation thereto.

4th. That Winans, in his specification, developed no new principles of mechanics, but merely borrowed all that is sensible in his philosophy from Allen, Jervis, Chapman and Tredgold.

5th. That his patent was issued immediately after the building of the "*Comet*," which was in no way,—excepting the shape of body,—different from the Columbus, and, that the "Washington Cars," which were the alleged result of his experiments, were built *after* the invention and construction of the Philadelphia car, "Victory," which was the predecessor of all the modern railroad cars.

6th. That Winans's claim has nothing to do with drawing by

the body, and he had nothing to do with the invention of that mode of draft.

7. That putting other men's machines to new uses, gives Winans no right to a patent; and especially, that using old running gear to carry one kind of a load rather than another, gives the Plaintiff no right to a patent.

8th. That in order to defeat a patent, it is *not* necessary to show that prior machines existed in every respect similar to the one patented; as a change of proportions will not support a patent.

It is not expected that the Court will be able to remember 600 pages of printed testimony, therefore I shall take no time in reading it. I can only refer to and make comments upon it, requesting your Honor to verify my references at your leisure.

The *conduct of Winans* from the time of his return from Europe,—where he had been trying to sell his friction wheel,—to the 1st of October, 1834, shows plainly that he did not dare to claim any invention of the running gear of the eight-wheel car.

1st. Consider his relations to Conduce Gatch,—the inventor.

2d. While in the Reports of the Engineers and others connected with the Baltimore and Ohio Railroad Company, mention is made of every little invention which any of them laid claim to, Winans is not mentioned in any part of the reports up to the 1st of October, 1834, in any particular, as the inventor of the eight-wheel car.

3d. That the newspapers of the day, so far as they are brought to the notice of the Court, do not mention Winans in connection with this invention, though he is mentioned in connection with the friction wheel and other inventions.

4th. The running gear going into use before his eyes, in all sorts of cars, without objection or claim.

5th. The probability that Winans was one of the parties from the South, who visited the Quincy Railway. See the affidavits of Bryant and others.

6th. Winans lying by until nearly the expiration of his patent, and making but one attempt to enforce it, and that unsuccessful. I mean one attempt at law. That is enough for my purpose on that point.

I now proceed, in the second place, to call your Honor's attention to some of the more important inventions which were prior to that of Mr. Winans.

CHAPMAN CARRIAGE.

Of these the first is the "*Chapman Carriage*," *described in the Repertory of Arts.* Vol. 24, in 1814.

Chapman's carriage, as described and proved, represents both the six-wheel carriage now used, and also the eight-wheel carriage now used.

It was to be drawn by a connection with the body.

The motive power, as well as the passengers and freight, might be contained in it. And the very same thing is now being done in Boston, where they have a steam omnibus.

It was an engine and passenger carriage, such as were used in England, and called steam passenger coaches.

The trucks were left free to conform to the curves of the road; the axles were kept parallel; the bearing points of the wheels a distance apart equal to the guage of the track.

The trucks had the rigid wheel frames, consisting of the three cross-pieces, and the two side-pieces.

The middle cross-piece bore the body on a transom centre, and side bearings, with friction rollers, and was connected by king bolts with the truck.

The trucks supported the body, as all swivelling running gear always supported the body, that is, at or near the ends of the body.

It was to run on railroads, and it combined every essential principle of construction, arrangement and action, to run easily and smoothly on curves and the straight track.

It is proved the same in principle as the eight-wheel cars now in use.

That, Sir, is the *construction* of Chapman, and I have now only to add a very few words in regard to the criticisms upon it.

Criticism 1st is: That it was a *steam carriage*, and that the double truck, swivelling under a platform, bore an engine instead of freight. I answer: That the kind of load borne on the inside of the body, is quite immaterial to the arrangement and connection of the running gear; and the specification provides that this Chapman car may be driven by steam power, placed within the body, or any other kind of power, placed inside or out, horse-power or any other.

The 2d Criticism is, That the body was not quite so long as Winans's passenger or freight car.

Winans's car was from twenty to twenty-four feet long, and Chapman's car must have been about the same length, being about double the length of the ordinary car that then existed; but mere change in length *cannot* be material.

The 3d Criticism is, That the wheels were not quite so near in each truck as in Winans's patent.

We say, that they are as near as the wheels of the Defendant *between the bearing points* thereof, which is the essential distance.

In other words: the distances of the axles in Chapman is as near as the Defendant's; that is, the material distance, and not the distance of the flanges. (See Byrne, p. 12.)

That there were other cars besides those which carried the engine—called the Chapman freight cars. The Plaintiff's criticisms had no application to *those* cars—as they were merely freight cars, and were called " Chapman's cars."

These, Sir, are all the objections that have been made. I think, however, if my memory serves me, that Mr. Keller suggested that there was no mode of drawing this car by the body shewn in the book, though there *was* a mode of drawing the six-wheel car by the body thus shewn.

I think Mr. Keller said that there was no mode of drawing *described*; if he did say so, he is mistaken, as he will see by looking at the book from which the patent is taken. If he did not say so, no further answer to that objection is necessary. The Chapman book shows cars as drawn by the body, and trains also, as drawn by the body. I think that the learned Counsel made another objection, which was, that this car was made for a tram-road, which I answer by saying that the description and the specification speaks of the *edge-rail*, which is the same thing as that which we call the T rail; and which we use at this day. The wheels had a ledge or flange, which shows that it ran upon rails of the same description as we use, and whether constructed in the T shape or any other, it is no matter.

I think that one further objection was made, and which was so unfounded that it almost escaped my notice,—that the wheels revolved *on* the axles, and *not with* the axes.

In the first place, we do not admit that the fact is so. But if it be so, it is quite immaterial to the principle of the Chapman car, because the revolving of wheels *with* their axes, was a well-known mechanical equivalent for revolving the wheels *on* their axes; and this fact would not change the combination of principles upon which the double-truck swivelling car is constructed.

And we further deny, that it is the true construction of the Chapman car, so far as relates to the wheels. And, inasmuch as the invention of Chapman has been repeated by other inventors over and over again before the patent, I do not perceive how any little supposed difference of that sort can be thought to be material.

TREDGOLD CAR.

I now proceed to the consideration of the "*Tredgold Car.*"

The mode of its construction and operation having been carefully examined, as it is shewn and described in the treatise, by many distinguished scientific and practical engineers and machinists, is proved by the testimony of the following witnesses, viz:—

1.	Conduce Gatch,	page 24.
2.	Oliver Byrne,	11, 62, 3, 4, 5.
3.	John C. A. Smith,	29.
4.	Isaac Knight,	32.
5.	Edward Martin,	36.
6.	George Beach,	39.
7.	Stephen W. Worden,	40.
8.	L. R. Sargent,	41, 2.
9.	John Wilkinson,	52, 3.

10.	George Law,	54.
11.	Jacob Schryack,	73, 4.
12.	Wm. T. Ragland,	93.
13.	Henry Waterman,	97.
14.	George W. Smith,	99.
15.	Charles B. Stuart,	111.
16.	Wm. J. McAlpine,	114.
17.	Vincent Blackburne,	117.
18.	Walter McQueen,	120.
19.	George S. Griggs,	126.
20.	Albert S. Adams,	128.
21.	James H. Anderson,	131.
22.	John B. Winslow,	133.
23.	Asahel Durgan,	136, 7.
24.	John Crombie,	139.
25.	Henry W. Farley,	128.
26.	Godfrey B. King,	163.
27.	John Edgar Thompson,	166.
28.	John Murphy,	175.
29.	William Pettit,	178.
30.	Asa Whitney,	179.
31.	William C. Young,	191.
32.	William P. Parrott,	199.
33.	Charles Minot,	202.

Now, what were the objects described by Tredgold, for which his eight-wheel car was designed?

1st. To carry great weights on railroads. (See p. 94.)

2d. To carry them upon such railroads as were then built,— on roads with such curvatures, grades and inequalities as then existed.

To accomplish this object, he recommends a plan, which is precisely the same as Chapman had patented; and the same as a double Newcastle coal wagon.

What were the conditions to which it was necessary that this carriage should be made to conform?

As to the roads, the *conditions* were:

1st. They were full of short curvatures.

2d. They had switches and turnouts.

3d. They had changes of grade.

4th. They had inequalities in the rails round every curve. (See p. 135.)

5th. Also, other irregularities, occasioned by local depressions or elevations of one or the other rail, both at the joints and in the general line of the road.

6th. I will now illustrate, by the model of the track, another of those inequalities in the level of the rails, particularly mentioned in Tredgold's book.

On all parts of the road where curves are constructed, the inner rail usually continues in the same plane throughout its entire length. The outer rail rises gradually from the beginning to the centre of the curve, and then descends again to the plane of the opposite rail.

Both these rails cannot be contained in the *same plane*, and the curves they make are therefore called curves of double curvature, the same as those mentioned in Tredgold on page 135, and which I will now read.

Then how did Tredgold make his structure, so as to conform to any of these irregularities and inequalities?

1st. As he must carry a heavy load, he began with a long body.

2d. As eight wheels fastened by their axle-trees *to the frame work* without swivelling would not pass the curves or switches, he placed the *then* old fashioned four-wheel swivelling bearing carriages under each end of the body.

3d. The draft is presumed to be the same as that of every other car shown in the book: namely by the body. (See plate 1.)

Two advantages were presented by this structure besides the carrying of great loads, viz:

1st. To distribute the weight equally on *all the eight wheels*.

2d. To distribute it over a considerable EXTENT of the road as much as was useful, besides swivelling and conforming to all the inequalities and curvatures of the track.

Now let us examine as to the manner and philosophy of this distributing the weight equally on all the eight WHEELS.

This object of making the *weight* borne on each one of the eight WHEELS EQUAL, we say is accomplished, whether the two trucks are near or remote, each wheel bearing *one-eighth* of the whole load, because all the load is borne on the CENTRE of the trucks, and each truck is equi-distant from the two ends of the body. (Mr. Whiting here explained his views by aid of the models, and then proceeded.) The point which I wish to explain is, that each wheel bears an equal portion of that load. Half the body being supported by each truck, and each truck (having four wheels) swivelling in the centre, each wheel must therefore bear one quarter of what each truck bears.

We will say that the whole load shall be 800 lbs. *Each truck* will bear 400 pounds, and each wheel will bear one-fourth part of what the whole trucks bear, viz: 100 lbs., and this equal bearing of the wheels results from the introduction of the swivelling principle of the trucks.

Well, each wheel bears an equal weight, whether the body be short, so that the eight wheels *need not be* long or *equi-distant* to accomplish that object. It would not be the case, that is, each wheel would not bear an equal weight, if the eight wheels were fixed to a body, *without swivelling* at all.

Suppose you make a long rigid car body with eight wheels, each pair turning on or with an axis which is unable to swivel, being fastened to the body like the rear axles of a road wagon; suppose that with such a car you undertake to pass round a curve. How would you succeed ? Could this be done in such a manner as directed by Tredgold, viz: so as that at all times, the *stress* of all the wheels upon the rails shall be EQUAL, so that one shall bear upon the rail with just as many pounds weight as the other seven wheels ?

With such a structure this is impossible. For as soon as such a car enters upon the curve, the bearing point of *one* of the two forward wheels is lifted above the plane which it has ascended, so that the weight of the load may be borne principally, if not wholly, on the front and rear wheels; while the intermediate ones bear little, if any, of the weight. So also whenever in passing on an even grade, there occurs a sinking of the rail for a distance less than the distance of the bearing points of the front and rear wheels; the intermediate ones bearing little, or none of the weight: unless the body of the car gives way.

The same inequality of stress occurs when there is a mere change of grade in the rails.

Furthermore, the same difficulties would occur, excepting the last above stated, if the car should have four-wheel trucks, which could not swivel VERTICALLY, but only HORIZONTALLY by means of an axis, like that shown in the model.

In the first place you could not pass a curve, and why? Because the wheel-frames remaining fixed, prevent the least lateral swivelling whatever, and therefore you cannot get round the curves, and when you come to a switch you could not pass in or out of a depot, because a switch is sharper usually than the curves on the road.

To state the proposition abstractly. Any inequality on the rail, which would tend to lift up one wheel, or throw the axes out of the same plane, it would prevent the equal bearing of the wheels on the rails at *that time*. Therefore, to pass any inequality in the rails, and distribute the weight equally on all the wheels at all times, the trucks must swivel laterally on a king-bolt.

Now, as neither of the structures just supposed would answer *any* of the purposes proposed by Tredgold, while the trucks swivelling on a centre pin or king-bolt *would* answer all these purposes, and as swivelling trucks had been familiar to the English engineers from 1812, when Chapman took out his patent, down to 1825, when Tredgold wrote, It is incredible that so distinguished a mechanician as he was should devise and recommend a car so constructed as to be incapable of answering one of the purposes for which it was devised.

The next *object* of Tredgold is, to distribute the bearing of the load over as large a reach of the rail as might be useful, and no farther.

Now what is the rule of utility and necessity that is *laid down in Tredgold?* It is, that the body must not be so short, and the wheels in each truck must not be so near as to bring two wheels, or the crushing force of two wheels at any one time BETWEEN TWO SLEEPERS, or, on *one* SPAN of the rail.

It is quite immaterial to this question whether a rail is nine

eet long or twenty feet long. It depends upon the sleepers for its
upport, and these are sometimes made of stone and sometimes of
imber.

The rule laid down by Tredgold is the sensible rule which is
now followed in the arrangement of the wheels, and which is, to
put them so far apart that two wheels shall not come together at
one time, with their crushing weight, on any one span of the rail.
And, after you have separated the wheels in each truck to that
extent, there is no use in spreading them any wider, *so far as
regards the strength of the rails.* The distance to which the two
trucks are separated is wholly immaterial, provided this rule above
stated is observed.

It is the using of the two swivelling trucks supporting the
whole load or body at two points, that makes each truck support one half the load; and each wheel in each truck supports
one quarter of the burden that the whole truck bears. The distance or nearness of the trucks does not change this law.

It has been supposed, that it is a fundamental principle of the
Tredgold car, to distribute the weight as equally as possible upon
the rails, and this is certainly done; but this *equal distribution
does not depend upon the equi-distance of the wheels.*

And it has also been supposed that this equal distribution of
the weight on the wheels, and through them on the rails, was incompatible or inconsistent with having a long body, and the weight
all borne by two trucks far distant from each other.

Now it has been demonstrated, that each wheel bears one
eighth of the load, wherever the load is placed. The present
eight-wheel cars are all so arranged as to distribute the weight
on the rail precisely as stated by Tredgold, viz.: that the wheels
are so far apart that two cannot rest at once on one span.

Judge Nelson. The stress of *all the wheels* on the rails, where
the bearing points are, will be *equal,* no doubt.

Mr. Whiting. (Presenting to the Court a diagram.) What
does Tredgold mean here, (in diagram,)—is it for distributing the
weight evenly upon the RAILS?

Now let Tredgold explain his own idea. (Referring to Plate
1.) He says, " This is a diagram to show how a wagon may be
made with eight wheels, so that the stress of each WHEEL upon

the RAILS may be made equal." It is an equality in the stress on *each wheel* which he desires to secure. There is no statement of Tredgold in which he says that his object is to do any thing else.

Is not the stress of each wheel on the rail always equal so long as you use a truck swivelling on a king-bolt? But the moment you put on *side bearings*, it is not so; and the moment you put on a truck that does not swivel, it is not so. It is the equalizing of the stress on each WHEEL that is the GREAT IDEA of Tredgold, in order that each wheel should ALWAYS support one eighth part of the load.

Mr. Hubbell. I will read a passage where he shows that clearly. It is on p. 95 of Tredgold. "The load on EACH WHEEL must be limited to suit the strength of the rails, and the *weight* is seldom to exceed *two tons*."

Mr. Whiting. Your Honor's desire is, to see if Tredgold's plan is *inconsistent* with the long body. Now, if it had not happened, merely as a convenient mode of drawing, that these wheels (referring to the diagram) had been placed about equidistant from each other—(which is a mere accident,) and not an essential feature in his car,—no question of this sort would have arisen.

To prove that there is nothing in Tredgold's plan inconsistent with substituting a long body instead of a shorter one, whereby the two trucks are placed farther apart and the eight wheels no longer equi-distant from each other, I point to the opinions of some forty scientific and practical men, who have carefully examined the subject, who *ought* to be competent to read and understand a treatise on machinery like Tredgold's, and who *ought* to be able to inform this Court what is and what is not consistent with Tredgold's plan, as described and shown in his work. And as these are questions not of law but of physical *science*, it is not too much to suppose that those who have devoted their lives to these subjects can understand them as well as we.

If Tredgold can be shown, by the gentlemen on the other side, or by anybody else, to say in direct language, or by fair implication, *that in order to answer the purposes for which his car is designed, it is necessary to place all the wheels equi-distant from each other along the line of the track, then I will give up the case.*

And, if they cannot find any such statement, and I can find expressions which are wholly consistent with placing the Tredgold trucks at the ends of a *long-bodied* car as well as a short-bodied car, then the length of the body is *wholly immaterial*.

We say that there is nothing in Tredgold which prescribes the distance of the wheels from one another, with one exception, and that is, that you shall not put two wheels on the same span of rail.

And there is nothing in Tredgold's book, or drawings, or in the idea or plan of his car, or in the uses to which it was intended to be put, that is in the slightest degree inconsistent with the substitution of a long instead of a short body, whereby the eight wheels would no longer be equi-distant from each other upon the rails.

I have already said that there is but one rule laid down by him as to the distance of the wheels from each other; that rule is to prevent the wheels from being, under any circumstances, too close together.

After stating, on page 29, the *length*, size and strength of the cast iron edge rail in actual use, and the proper mode of supporting the rails on sleepers of *stone or wood*, *solidly* embedded, and of great weight, he proceeds, on page 133, to show what is the best *length*, as well as the best shape of rails. And he says, "Now it is obvious that if ANY support be incapable of sustaining half the weight of the wagon without injury, it is insufficient for this purpose; consequently, if the supports be ever so numerous, the same degree of firmness becomes as necessary as if they were at a *great distance apart*." See also page 132, for the best manner of *embedding* these unyielding supports or sleepers. *Each sleeper* being thus sufficient to support a greater weight than can ever be placed upon it by the stress of any *one wheel*, and the wheels being placed so far apart that only one wheel can rest at one time on any one *span of rail*.

The Court will notice that the STRESS on the RAILS, is, in all cases, equal to the stress on THE WHEELS.

To equalize the stress on each wheel will equalize the stress of the wheels on the rails, so that ONE bearing point on any part of the rails will never have more weight or stress upon it than any of the *seven* other bearing points must bear at the same time.

And the great idea of Tredgold was, that by making his double truck swivelling carriage, he should always make the number of TONS WEIGHT borne upon one of the *eight wheels*, and by that wheel upon the rail, equal to one eighth of the whole load, and thus equal to the number of tons weight on either of the other *seven* bearing points, wherever they should happen to be placed.

I will now read to the Court the following extracts from Tredgold's Treatise to prove that this is his plan, and that it is wholly without reference to the equi-distance of the wheels.

It will be remembered that in constructing passenger cars, he recommends (see page 93) the application of springs. The first passage to which I refer is in chapter 5, page 94, in Tredgold, where he says: "Small carriages must obviously be both heavier and more expensive, in proportion, than large ones. But as the *stress on a wheel* must be limited on a railroad, we cannot much enlarge the carriages without adding to the number of wheels."

On page 94, the subject of the author is eight-wheel carriages. He says: "When a carriage has more than four wheels, the body must be sustained so that its pressure may be divided equally among the wheels. In the case where eight wheels are applied to support one body, if the body rests upon the wheel-frame of each set of four wheels, in the middle of its length, (see Fig. 26, Plate 4,) and is connected with those frames so as to allow the greatest possible change of level on the rails, it is obvious that each wheel must bear an equal pressure. If one frame with its four wheels be removed, and an axis with two wheels applied in its place, the carriage would have six wheels, and it would be easy to adjust the load so that the pressure on each pair of wheels would be equal."

The next is at the end of page 94 and beginning of page 95, where it is said, "The load on *each wheel* must be limited to suit the strength of the rails; it will seldom exceed two tons *on a wheel*, nor be less than half a ton. The size of the axles may therefore vary from 2·2 inches to 3·5 inches. Perhaps the most advantageous load will be about one and a fourth tons on each *wheel*, which will require an axis of three inches in diameter."

The next is on page 179, Plate 4, Fig. 26. "A diagram to show how a wagon may be made with eight wheels, so that the stress of *each wheel* on the rails of a Railroad may be *equal*. The body of the wagon rests on the wheel-frames at A. A., and is connected to them by an axis on which the frames turn, when, from any inequality, the axes of the wheels are not in the same plane. See p. 94."

The next is on page 101. "In proportioning the body of a carriage, it should be kept in view that the load should be as low as possible, and particularly where the inclined planes are steep, for a high load in such cases produces a very unequal stress *upon the wheels*, and *consequently* upon the rails."

The next is on page 173. "For carriages on springs, and steam carriages, consider the stress on each wheel only two-thirds of the actual load upon it, which will be about an equivalent excess of strength for this case."

The next commences on page 135. "When a considerable degree of curvature is given to a railroad, the rails of the outer curve should have a slight rise to the middle of the curve, and the rails should be stronger in a lateral direction in both lines. The object of making a slight ascent to the middle of the curve of the outer rail, is to counteract the tendency of the carriage to proceed in a straight direction, without its rubbing so forcibly against the guides as we have observed in cases where roads have had a considerable curvature."

Next on page 126. "The distance between the wheels of the carriages should be such that the *unsupported part of a rail should have to carry only one wheel*."

Next, commencing on page 12. "In some parts near the staiths we observed malleable iron rails, in fifteen feet lengths, supported at every three feet. (See Figs. 9 and 10.) They are three and a half inches deep in the middle between the supports, and two and one-fourth inches in breadth at the upper surface; one yard in length weighs about twenty-eight pounds.

The wheels of the coal-wagons are 2 feet 11 inches in diameter, with ten spokes, and weigh $2\frac{3}{4}$ cwt.; and their axles are 3 inches in diameter, and revolve in fixed bushes."

And again on page 13. "The wheels of the engine carriage are 3 feet 2 inches diameter, with twelve spokes in each, and each weighs $3\frac{3}{4}$ cwt."

Next on page 31. "The length of each rail being 9 feet, it is supported at every three feet, and is $2\frac{1}{4}$ inches deep, and $\frac{3}{4}$ of an inch thick; the wagons carry about 35 cwt."

Next on page 43. "It often happens that a great part of the resistance at the rails arises from the lateral rubbing of the guides of the wheels; therefore it is desirable to give the wheels a tendency to keep in their path, with as little assistance from the guides as possible.

"For edge-rail carriages this may be accomplished by making the rims of the wheels slightly conical, or rather curved, as shown in Fig. 24; the carriage will then return of itself to its proper position on the rails, if it be disturbed from it by any irregularity."

Next on page 93. "Carriages for common railroads are made strong, to resist the shocks they are exposed to at every change of velocity; and it is necessary to make the parts which come in contact solid pieces, extending a little more than the length of the body of the carriage, and hooped at the extremities to prevent splitting. But carriages for passengers and for various kinds of goods must be provided with springs to reduce the force of these shocks."

Not *one word is said* in the whole book, as to there being any necessity to place the bearing points of the wheels at *equal* distances from each other along the line of the rail.

And now I will read the lines opposite to Plate 4, on page 179. "A diagram to show how a wagon may be made with eight wheels, so that the stress of each wheel on the rails of a railroad may be equal. The body of the wagon rests on the wheel frames at A A, and is connected to them by an axis on which the frames turn, when, from any inequality, the axes of the wheels are not in the same plane."

Of course the stress must be on the *rails*. Therefore, in the four or five times in which this matter is introduced, in every instance he has mentioned that it is among the WHEELS that the weight is to be divided, and not a distribution of the load at equidistant points along any given length of the rails.

It is obvious that NOTHING is gained in the way of distributing the weight along a greater or less extent of rails, provided each wheel has one span of rail to itself at all times.

And as one wheel cannot rest but on *one span* at one time, it is quite immaterial whether the spans on which the wheels respectively rest, follow each other in consecutive order; so that the first wheel stands on span No. 1, the second on span No. 2, the third on span No. 3, and the fourth on span No. 4; or whether the wheel No. 3 stands on span No. 20, and wheel No. 4 on span No. 21. Because *each span* bears one *wheel;* each wheel bears one *eighth part* of the load. And therefore it is obviously of no importance whether the wheels are equi-distant or not. Following Tredgold's rule—no position of the wheels can be imagined in which *one span* of rail can support more or *less* than one wheel, or one eighth of the load, whether these spans are consecutive or ever so far apart.

If I understand rightly, Mr. Keller contends that the Tredgold car did not have an axis or king-bolt, that would swivel laterally, but that the axis would only swivel vertically.

Mr. Keller. I said, whether Tredgold had a vertical or horizontal axis, so long as the wheels were upon axes permanently connected with the frame, and no springs were interposed, it was impossible that the wheels could bear equally upon the rails, any more than a chair with unequal legs could bear equally upon the floor.

Mr. Whiting. This remark shows that the car must have had an axis swivelling laterally; otherwise it could not conform to *the curves,* bear equally on the rail, or answer any one of the objects for which it was made; and that the connection between the car and the truck could be *permanent* in no other sense than is applicable to the ordinary connection of the truck to the body by the king-bolt. I understood another objection to be, that if this car in Tredgold had what we call a vertical axis, or pivot, it could not rise up grade. To which I simply answer, that all modern cars which have the king-bolt, do rise up grade, and go round inequalities without difficulty.

And in regard to the necessity for springs; the result of experience is, that there is no necessity of springs for that purpose; and this is shown by the Quincy cars, which carried some sixty tons up steep inclined planes without springs, and swivelled round curves, and did not require them.

Another consideration is, that as Tredgold was a man of science and not a sciolist, and as he had Chapman's book before him, with a *wheel frame*, and cars with swivelling trucks pictured as standing and swivelling on the curves of a railroad in a situation that demonstrated the impossibility of having a long body with eight wheels without swivelling, can it be possible that he could have been so stupid as to recommend such a car as the Counsel for the Plaintiff would attribute to him?

To put an end to such a pretence, I will refer to Plate 5, Vol. 24, (*Chapman's Patent*,) p. 136 of the Repertory of Arts, also pp. 130 and 139.

Is it possible, that having that work before him, that so eminent a scientific engineer as Tredgold, who received a pension from the English government for his public services, should have stultified himself so as to have made such a miserable boy's toy, as the learned Counsel on the opposite side would have your Honor understand it?

But suppose, for the sake of argument, that the design of Tredgold was to have the wheels of his carriage placed at equal distances apart, along the rail, and that he had not thought of the fact that by doubling or quadrupling the length of the body (the trucks remaining the same), the car would run more steadily on the track.

Can it be said, that any one who first put a longer body in place of a shorter one, can have a patent for such a mere alteration of form or proportions?

How absurd for the Plaintiff to set up a pretence that making the body longer or shorter introduces into Tredgold any *new prinple*, when he himself states the CONTRARY in substance. He says in his specification, page 5, fol. 20, " When the bolsters of the bearing carriages are placed under the extreme ends of the body, the relief from shocks and concussions and from lateral vibrations is greater than when the bolsters are placed between the middle and the ends of the body, and the relief is not *materially varied by increasing or diminishing* THE LENGTH *of the body*," &c.

Again. As we find in Tredgold a double truck swivelling eight-wheel car, having the *capacity*, when put to use, of accomplishing all the objects for which the Defendants construct their cars, viz., swivelling to the curves, travelling smoothly and evenly over the curves and inequalities of the roads, it is of NO CONSEQUENCE what was the OBJECT or DESIGN for which Tredgold's car was made. If the thing made, answers our purpose, that is sufficient to protect us. No other person can maintain a patent for the old car, because he has found out that it has more advantages in practical use than the original inventor knew of.

If Tredgold were now living, he could not be enjoined from using his own invention, with a longer or shorter body, because a subsequent party had discovered that it would carry its load still more easily if made of greater length. Tredgold would have had a right to his own invention and to all those CHANGES of *proportion* which were *incidental* to the use of this, as well as of all other machinery.

And I now proceed to the Quincy, Allen, Baltimore trussell and wood cars, and other inventions prior to the Plaintiff's.

But in order to do justice to our case, it is necessary, before approaching those topics, to draw the attention of the Court to some extraordinary doctrines that have been put forth by the other side.

One of the notions I refer to is, that if Winans had seen an eight-wheel car in all respects perfect, except that it drew by the truck, and that he then devised a mode of drawing it by the *body*, he would be entitled to a patent for the car as a whole.

That, Sir, is strange law, and your Honor will observe that counsel so able to present their case in its best aspects, would not have taken such a position unless they were driven to it by stress of weather.

MODE OF TRACTION.

Another of the Plaintiff's *notions* is that Winans discovered a *new idea*, viz: that drawing the car by the body, instead of by the perch, would allow the trucks absolute freedom of swivelling! In answer to that I say, that the same mode of traction by a coupling from the middle of the ends of the body was shown in Chapman's patent in 1814; in Tredgold's works in 1825; and in Strickland's in 1826.

These two modes of drawing therefore, by the perch, and by the body, in cars with swivelling trucks, were well known, and in public use before the date of the patent, though not adopted by

Winans. Also drawing by the king-bolt, which is substantially the same thing as drawing by the body, was in public use in 1833 and 1834, on the New Castle and Frenchtown Railroad, and on the Philadelphia, Wilmington and Baltimore Railroads. (See Dorsey, Defendants' No. 4, p. 77, folio 293.) We also say, that the Allen engine embodied the same idea, as the power that propelled the trucks was delivered to the trucks through the king-bolt, leaving the trucks free to swivel.

The sixteen-wheel Quincy cars had three trucks swivelling under them, and the draft was by one of the truck frames. When the cars are to be drawn singly or by horse-power, it is very doubtful which way is best. Each has its advantage. But when cars are to work in trains, drawing by the body is the most convenient. We say that, in point of fact, the mode of drawing by the middle of the end of the body, was invented and introduced first on the Baltimore and Ohio Railroad by Jacob Rupp, in February or March, 1835, after the date of Winans's patent, on the 110 freight cars, which were built by him for that road. And it was while they were building, that the idea of drawing the Washington cars, by the body, was first suggested. On this point I will refer to

  1. Rupp,   Defendants' No. 2, p. 17.
  2. Shryack,   " " 3, p. 40.
  3. Gatch,   " " 3, p. 27.
  4. May,   "
  5. Forest,   " " 2, p. 19.
  6. Shultz,   " " 3, p. 43.

The Car "Victory," which was invented by George Fultz in Philadelphia, *modelled in* 1829 *or* 1830, and commenced in September of 1834, although not in operation until July 4, 1835, had a draft by the middle of the end of the body, and was 35 feet long, and 'square in form.

Moreover there is no description in the specification of any mode of draft, nor is there reference made in the patent to any drawing by which any mode of traction is shown.

There is no intimation, that the body of the car was to be drawn otherwise than by the perch, although both ways had been tried and were well known.

There is no claim in the patent relating or alluding to any mode of drawing.

Mr. Keller says, that it is absurd to suppose that Winans did not intend to embody this great principle of drawing by the body, when this was the great fact that gave vitality to the whole structure.

Then let me ask, why did they build the "Comet" to draw by the perch, after this wonderful discovery had been made, and had already been embodied in the Winchester, as the Plaintiff pretends it was?

Why did Winans not mention so important an idea in his patent, which is so full of ideas that are not important?

Why did he not claim it? Surely the persons who drew the claim were shrewd enough.

If there was any drawing originally sent to the Patent office, why does not Winans produce a copy of it?

Did he never have it?

What was the mode of traction of the car as shown in that drawing?

Had he no model?

Why does Winans not swear in his affidavit, that he applied, or intended to apply, that mode of drawing, if that statement is true?

We have shown that the picture of the freight car, sent in 1837 to the patent office, does not aid the specification, but is directly at variance with its directions.

If there was any grand idea which gave vitality to the whole invention, why did not any body get hold of it? Why was it not expressed by Winans himself, or by Dr. Jones, or Mr. Latrobe, his counsel?

Why was it not alluded to by some one of those who had a hand in making Winans's specification?

Since drawing by the perch had been the common way, why leave people in the dark, if any new way was to be substituted?

Why is not any allusion made in the patent to the subject of drawing *trains* of cars together, a subject so much dwelt on by the Plaintiff's counsel?

What was all the alleged experimenting for, if the result be not stated?

The perch prevented the trucks from swivelling too much or bobbing from one side to the other, when the wheels were close together. Drawing by the body became necessary and was introduced generally when the wheels of each truck were spread wider apart.

But when the rails alone were relied on, to keep the truck square, the perch being dispensed with, the wheels in each truck were placed further apart for safety.

And if the mode of drawing be essential to the principle of Winans's pretended invention, it not being stated, makes his specification defective in a vital part.

If I were to come before your Honor, expecting to maintain a patent, and if the mode of drawing was a material and substantial element in the invention, and if it had not been described or alluded to in the specification or claim, I should expect to be told that my patent could not be sustained, and that it would be absolutely necessary for me to get it re-issued. No court could sustain a suit, where such a material element was not alluded to, described or claimed.

Chief Justice Taney has decided that the mode of drawing constitutes no part of Winans's claim, whether he did or did not invent it.

Therefore, the more importance the Plaintiff attributes to this feature, the worse for him. Judge Conkling says the same thing in Plaintiff's proofs, p. 7, as follows:—" If, indeed, the infringement complained of had consisted in the use by the Defendants of this new mode of traction, the action I think could not be maintained, for I am of opinion, that according to the true construction of the specification, the Plaintiff's claim does not extend to this mode of coupling, it not being mentioned at all in the written specification, and his claim being ' the before described manner of arranging and connecting the eight wheels,' &c., he has limited himself to what he had before described; nor do I understand him now to claim any thing beyond this."

And that opinion was given, notwithstanding the Judge was under a mistaken impression as to the extent to which the drawing annexed to the Letters Patent was to be used, for he thought that it was to be taken as *part* of the original patent.

There is not even a suggestion relating to a *train* of cars in his specification, and the patent is to be treated as though only one passenger car were to be used at one time, as had been usually the case.

Who will believe that Winans had any knowledge as to the importance of drawing by the body in trains, when he is so particular as to other things, but never mentions this?

The Plaintiff's counsel has admitted that there is no infringement, if we use all other things constituting Winans's car, except drawing by the body; and it has been judicially settled that this constituted no part of the Plaintiff's invention or claim.

The car as described by the Plaintiff's patent was to draw, *not by the body, but by* the perch, and the Plaintiff says, in express terms, that " the end which I have in view may, nevertheless, be obtained by constructing the bearing carriages in any of the modes usually practised, provided that the fore and hind wheels of each of them be placed very near together." Plaintiff's proofs, p. 174, fol. 14.

The Plaintiff's counsel says that this refers to the bearing carriages on the Baltimore and Ohio Railroad; but as the specification does not describe them, none of Defendants' experts knew how they were built. But the evidence now offered at this trial shows how the bearing carriages were constructed.

They were *all drawn by the perch*, and I will refer to the following of the Plaintiff's witnesses on this point.

1. John Elgar, Plaintiff's proofs, p. 27, fol. 116.
2. Wm. Woodville, " " " 32, " 140.
3. Michael M. Glenn, " " " 37, " 165.
   " " " " " 38, " 170.
   " " " " " 39, " 174.
4. Oliver Cromwell, " " " 46, 209, 212.
5. Thomas Walmsley, " " " 58, fol. 250.
6. John Ferry, Plaintiff's proofs, p. 58, fol. 269, 270—272.

All the Defendants' witnesses agree, that all the " ordinary bearing carriages " drew by the perch, and by the perch alone.

Now the Plaintiff has suggested one alteration of the ordinary bearing carriage, in order to suit it completely to the requisitions of his patent ; that is, " to bring the wheels very near together ;" and no other alteration being suggested, " *expressio unius, exclusio alterius.*"

To secure the alleged new idea, of the entire freedom of the swivelling of the trucks under the body, (if Winans had conceived such an idea,) his plan, described in his specification, is to bring the axles of the wheels as near together as possible, so as to resemble the action of a " single wheel," and not by the mode of drawing by the body, as the Plaintiff's counsel contend.

And to make such trucks safe or practical,—owing to their tendency to swivel too much,—it is necessary that this tendency should be controlled by a *perch*. Now, as Winans says that all the advantages claimed by him could be enjoyed by using the common bearing carriage, which drew by the perch, it amounts to the same thing as saying that none of the advantages contemplated by him, were lost or impaired by drawing the trucks by the perch.

The free swivelling of the forward truck was not intended, for it would have been dangerous.

To show the reason of this, I refer to the testimony of Godfrey B. King. Defendants', No. 3, p. 81, fol. 308—9.

Now consider the admission of counsel, that you may take every thing else in Winans's patent, if you do not draw by the body, and you do not infringe ; then, clearly, you may take the *very* car described by Winans and not infringe.

The true principle of the Winans car, according to the counsel's version, is not embodied in any car that does not draw by the body.

According to the Plaintiff's testimony, the difficulty that occasioned running off the track, was not by drawing by the perch, but by having the perch too long, so that the bodies of the cars did not come together, when stopped, as they now do.

If the perch were too long, it must come in contact with the engine or other cars when the train stopped.

The idea, that connecting by the perch was more dangerous in practice than connecting by the body, is erroneous in theory and untrue in practice.

The draft, in both cases, proceeds from the same point, and is applied at the king-bolt.

On the other hand, if the carriages were united by a rear perch and a front perch, the draft would be still in accordance with the line of forward motion, because it would be nearer to the cord of the arc in which the cars were passing, instead of a tangent to that curve.

To prove, further, that the idea of drawing by the body was no part of the plan of the Plaintiff at the date of his patent, we shall show that the Comet was the last and most perfect of all the preceding cars, and that the patent describes particularly that car, and that that car drew by the *perch*. And we do so by referring to

1st. Oliver Cromwell. Troy case, p. 26. He says that " the Winchester was next to the Columbus, the next was the Dromedary. I do not recollect any other car being built while I was there. My recollection is not distinct as to the Comet." His memory grew better in 1853. He then swears, at p. 48, that " the Comet was the fourth and last car previous to the 1st of October, 1834." Fol. 220.

2d. Michael Glenn, in the Troy case, p. 28, and Plaintiff's proofs, p. 40, fol. 182. He says that " the Comet was built last."

3d. In Judge Conkling's charge, p. 11, he states the same thing.

4th. Judge Taney's charge, p. 22, he is not quite so distinct. He states that the Winchester was built in March, 1834, the Dromedary in August, 1834, and the Comet came next before Oct. 1st, 1834.

Mr. Latrobe. We admit the fact as you have stated it, as it may save time.

Mr. Whiting. The admission will save time. Now, the drawing shown by the Plaintiff, of the " Comet," corresponds with the description of the patent, particularly as to one spring, and also in various other respects; but I will simply say, no other one of the cars preceding or following it corresponded, in its peculiarities, with the descriptions of the patent.

Now, how was the Comet drawn? I shall show, by the following witnesses, that it was drawn by the perch.

In Defendants' No. 2, Jacob Rupp says, p. 16, fol. 64—5, "the Dromedary and Comet drew by the perch."

In Defendants' No. 2, Leonard Forest says, p. 19, fol. 76—7, "the Dromedary and Comet drew by the perch."

In Defendants' No. 2, Conduce Gatch says, on p. 22, fol. 87, "that the Comet drew by the perch."

These affidavits have been more than a year before the Court, and none of the Plaintiff's witnesses have sworn that the Comet, which was the last of the four, drew by the body. In fact it is not alleged.

All the Defendants' witnesses unite in this testimony, and, therefore, the Plaintiff's experiments resulted in abandoning the draft by the body,—if he ever used it,—and this accounts for his silence on the subject of draft; and there is an end of all the philosophy on that point, so much talked of by counsel in this case; so much boasted of; so elaborately displayed. It was not thought of, described, alluded to or claimed by Winans or any of his able counsel who drafted the patent; and the whole theory, that Winans discovered the great importance of drawing by the body, and covered in his patent only such cars as *do* draw by the body, is wholly *without foundation*.

Judge Nelson. It cannot be necessary, Mr. Whiting, to go further in your argument upon this point; nothing is clearer to my mind than that Winans's patent has nothing to do with the mode of drawing the cars. He has not alluded to it in any part of the specification.

Mr. Whiting. I should not have taken so much pains to overthrow the arguments of my learned opponent, if he had not rested his whole case upon the hypothesis, that drawing by the body was an essential element in the invention of Winans, and that without this the invention was worthless and the patent practically good for nothing; and if the counsel, who thus staked the whole fortunes of the case upon this assumption had not spent nearly half a day in trying to satisfy your Honor that he was right.

Judge Nelson. The argument made no impression whatever on my mind. The patent has nothing to do with the mode of drawing.

Mr. Whiting. I will then pass on to my next point.

THE WASHINGTON CAR.

It has been asserted more than once, that the Plaintiff's patent was founded upon improvements which were not developed until the building of the *Washington Cars,* so called. That until these were built, all preceding cars were merely a series of unsatisfactory experiments; that when these cars were planned and built, the experiments completed, and the philosophy developed, the specification was drawn up, and was intended to embody the principles of *those cars* as contra-distinguished from all that had preceded them.

Now we aver that these assertions are *totally untrue.*

The Washington cars were not planned before the date of Winans's application for his patent, and they cannot, therefore, be considered as a result arrived at before that application.

I have also shown that the patentee described, and intended to describe, the peculiar organization embodied in the "Comet."

To prove that the Washington cars were *not* planned before the specification was drawn up, the Plaintiff relies on these witnesses:

1st. Philip E. Thomas, p. 17, Plaintiff's proofs, fol. 68—9, who says; "that when the 'Washington cars' were about to be built, the plan of said cars was made a subject of much consideration and discussion, and the passenger cars for the said Washington Branch were directed to be built on the plan the Plaintiff recommended, *and which is the plan of those* AT PRESENT IN USE. *on said road;* so far as this affirmant understands and believes, that the Washington Branch was completed in 1835, and that the consideration and discussion above referred to, must have taken place some time previous, though this affirmant is not now able to give the exact date of it."

Mr. Thomas, therefore, does not fix the date as having been before September or October, 1834.

2d. George Brown, pages 21, 22; Plaintiff's proofs, folio 87 and 91 says, that at the meeting of the board of directors, which was held on the first day of September, 1834, the superintendent of transportation having charge of the carriages and moveable machinery on the road, suggested to the board of directors the expediency of providing the carriages and the locomotives to be used on the Washington Railroad. That the President in consequence of the said suggestion, was authorized *to submit a plan* of carriages of the most approved form for consideration.

Again he says at folio 89, "that the said car or carriage invented *and patented* by the Plaintiff was recommended and adopted by the board of directors." But he does not say, at what meeting, " and which was *afterwards* built," not *before* or at the time, and " put upon said road for use." If the car thus recommended *had been patented*, it is clear that the car could not have been *any other* than what was described *in the patent*, and the recommendation could not have been made until after the patent came into existence, viz. Oct. 1, 1834.

3d. John Elgar, Plaintiff's proofs, p. 25, folio 108. He says, " and that this improved car" (namely, the one described in Winans's specification) " was first brought into general use at the opening of the Baltimore and Ohio Railroad, on the 1st July, 1835," and not on, or before October, 1834.

4th. Michæl M. Glenn, on p. 41, folio 185. Plaintiff's proof says, " that several of said cars were completed and ready for use in the latter part of the year 1834. That there were ten of the said Washington cars built before the said road was opened on the 1st July, 1835."

He does not state that any were begun before the date of the patent.

At p. 42, folio 192, he says, " that he has been familiar with the Washington cars, so called, since the year 1835, in the fall of said year."

He does not fix the time in the fall, but it is certain that it was after the first of October, the date of the Plaintiff's patent.

5th. Oliver Cromwell, Defendants' proofs, p. 49, fol. 223—224, does not fix the time at all.

6th. George Maxwell, p. 62—63, does not fix the time except in 1833—4.

Thus far not one of the Plaintiff's witnesses asserts, that these Washington cars were planned previous to October 1, 1834, the time when the patent was issued.

6th. Mr. Latrobe himself fixes the time when the drawing of

the Washington cars was first shown him by Winans; which was after the 7th of February, 1835. (See Troy case Plaintiff's proofs, p. 26, ans. to Int. 7.)

These are all the Plaintiff's witnesses. The Defendant's witnesses on this point are:

1st. Jacob Shryack, Defendants' No. 3, fol. 156, says, "We commenced to build the Washington cars in the early part of November, 1834, as soon as the benches and shop at Charles street were prepared for the hands to work on them." He also states, "I am the foreman of the building of passenger cars of the Baltimore and Ohio Railroad Co." Again he says, "on the 28th October, 1834" (which was after the date of Plaintiff's patent) I went into the employment of the Baltimore and Ohio Railroad Co., and have been in their employ ever since." So that the cars could not have been begun before the 1st October 1834.

2d. Henry Shultz, Defendants' No. 4, p. 43, fol. 156, says that he is the "railroad car builder for the Baltimore and Ohio Railroad Co." He says further, "that the Washington cars were finished in 1835, to draw by the body, and that the idea in so doing was taken from the 110 freight cars made by Rupp and himself," fol. 165.

The car Victory (which was commenced in Philadelphia in September, 1834, and which came before anything was said or done about the Washington cars,) drew by the body.

No *plan was produced* for the Washington cars, because the Plaintiff's witnesses say, that the Railroad Co. adopted Winans's plan *as described in his patent.*

If any *plan* had been produced, why was not that plan inserted in the restored patent of 1837?

The Washington cars were in use in Baltimore in 1837, but the Plaintiff did not send to the Patent office a drawing of *any of them*, but substituted a *freight* car, to illustrate his description of a passenger car.

This fact, in connection with what was said yesterday, is conclusive that the idea of drawing cars by the middle of the end of the body did not exist in Winans's mind as a desirable arrangement, at the date of his patent.

Another proof that at the date of the patent, and even in 1837, when the drawing of the aforesaid freight car was sent to the Patent office, showing that Winans did not consider the mode of drawing as any part of his system, is the fact that he did not allude to it *on* the drawing, by lettering or describing it, as he did other parts upon the drawing which he did deem important. If he had done so, he could not have enlarged his patent.

To add still further confirmation to the proof, that the " Comet " was the foundation of the patent and that it was the only *result* of Winans's experiments, and that the Washington cars were not a result to which he had arrived before the specification was made, I recall the facts:

1st. That the " Comet " *was being built* in August, 1834, the " Dromedary " having previously been put on the road in August, the same month.

2d. *When* was the specification made? The invention of Winans's must have been completed before the specification was drawn.

The specification was drawn by Winans himself and by Dr. Jones, after which it was submitted to Mr. Latrobe.

Mr. Latrobe, in the Troy case, p. 24, Cross Int. 1, and at p. 26, Cross Int. 8, says, " that the specification was *on hand in his office* a long time; from six months to a year." He could not fix the time.

The patent was issued actually in October, 1834. The specification must therefore have been drawn six or twelve months before, that is, as early as April, 1834, if not as early as October, 1833, and having embodied what Winans then knew, and being the same specification now constituting a part of his patent; it follows that Winans could have embodied in that specification only such results as he had arrived at, at the time he wrote; and as it was written many months before any new style cars were planned or ever thought of, it is absurd to pretend that the Washington cars were the LAST in a series of experiments, which were never successful until these cars were produced.

No experiment could have been tried on cars that did not exist at the date of the patent.

No experiment could have been tried on cars, which were not planned by any person till between six and twelve months after the whole invention of Winans, as now shown in his patent, had been DESCRIBED and shown in his own hand writing in the specification.

Therefore in conclusion; the Plaintiff's patent was founded on such discoveries as he had made and embodied when the Comet was built.

It did not embrace the improvement of drawing by the body, subsequently introduced in 1835, into the Washington cars.

And if the validity of the patent is to depend on the Plaintiff's alleged discovery of the great importance of the mode of drawing, the fact is, that he had not made it, and he did not and does not claim it, but he expressly says in the patent, that which is inconsistent with such a pretence.

These directions in the patent, which are inconsistent with the absolute swivelling of trucks as derived from the drawing by the body are:

1st. That the bearing carriages must have the wheels very close together, so as to resemble the action of a single wheel (which makes them liable to wabble and turn round the track) and are to swivel under the bolster, in the *manner* of a common *road wagon*.

2d. The two trucks—after the wheels in each are coupled together by a spring—are also not merely placed under the body, or at or beyond the ends, but are " *coupled together.*" This is the language of the patent. The reason for it is, to prevent too great turning on the track, of the hind truck.

3d. All the advantages that Winans claims, he says may be derived by using, instead of the spring-trucks, any of the ordinary bearing carriages, if the wheels were near together, and all of which bearing carriages drew by *the perch* alone.

So that the Plaintiff had not planned the Washington cars, when he had completed his alleged experiments, and was not aware of any advantage to be derived by drawing the cars by the body.

These cars were *not the result* of any experiments arrived at before the specification was drawn. They are *not* the cars *described*, or intended to be described in the specification, and Winans had *not* in his mind the *idea* or plan of allowing the trucks entire freedom of swivelling, by arranging the draft through the body. But his plan *was* to have the draft through the perch, as in the Comet, which, as I have shown, was the last car begun or built before the *patent;* and *while* the specification was being prepared.

PUTTING AN OLD MACHINE TO A NEW USE.

Another question, arising out of a consideration of the history of prior inventions is—how far a change in the use or application of the running gear of a car for freight, to a car for passengers, or to a locomotive, would justify a patent ?

Indeed, it is well settled, that putting an old machine to an entirely new use, gives *no* right to a patent; and much less does putting it to an analogous use, *e. g.*, converting the running apparatus under a steam engine, or freight car, or passenger car, from one of these uses to the other.

The cases in point are :

Winans vs. The Boston and Providence Railroad Company, 2 Story's Rep. 412, in which the patent friction-bearings, which the Plaintiff claimed, had been applied to other carriages, but not to railroads.

In this case the Court held, that applying an old invention to a new use was not patentable.

Bean vs. Smallwood, 2 Story, 411.

In which the patent was for making the seat and stool of a chair in two parts, so that while the stool remains stationary, the seat is made to rock, &c., and enabling it to remain fixed in any angle, &c.

A similar apparatus had been applied, not to chairs, but to other machines, and the patent was held to be void.

"If this be so," says Judge Story, "then the invention is not new, but at most an old invention, or apparatus, or machinery, applied to a new purpose. Now I take it to be clear, that a machine, or apparatus, or other mechanical contrivance, in order to give the party a claim to a patent therefor, must in itself be substantially new. If it is old, and well known, and applied only to a new purpose, that does not make it patentable. A coffee-mill, applied for the first time to grind oats, or corn, or mustard, would not give a title to a patent for the machine. A cotton gin, applied without alteration, to clean hemp, would not give a title to a patent for the gin as new.

"A loom to weave cotton yarn, would not, if unaltered, become a patentable machine as a new invention, by first applying it to weave woollen yarn. A steam-engine, if ordinarily applied to turn a grist-mill, would not entitle a party to a patent to it, if it were first applied by him to turn the main wheel of a cotton factory. In short, *the machine must be new, not merely the purpose to which it is applied.* A purpose is not patentable, but the machine only, if new, by which it is to be accomplished. In other words, the thing itself which is patented must be new, and not the mere application of it to a new purpose or object."

No patent can be obtained for turning an old machine to a new use.

The *Jacquard power-loom*, for weaving figured silks, &c., could not be patented, because applied, without alteration, to weaving three-ply figured carpets.

A stocking machine would not be patentable, merely because applied to knitting seamless bags.

Blanchard's machine for turning gun-stocks would not be the subject of a patent, by reason of its being applied to turning lasts.

Neither would Goodyear's patent for vulcanized India rubber garments, if applied without any change of ingredients or of processes, to make car-springs, combs, or walking-sticks.

Nor the galvanic battery; because applied to the purpose of covering over a surface with silver or gold; or to the still more beautiful process of electrotyping.

Nor the electric telegraph, because applied to the new and admirable astronomical clocks, so as to obtain the difference of longitude between two places.

Nor watch-work, because applied to a sub-marine torpedo, to cause it to explode at a given time.

Nor a percussion lock of a pocket pistol, because applied to a large pistol, gun or cannon.

Nor any kind of axles formerly applied to common road wagons, because applied to railroad cars, or vice versa.

Nor whatever machinery has already been applied to railroad cars, by making it run faster, or carry one species of burden rather than another.

Winans himself knew that it made *no difference* whether his double truck was put under a locomotive or a passenger car, as he sold the right to use it under locomotives to John Tucker, for the Philadelphia and Reading Railroad, in November, 1845.

See Tucker's affidavit, Defendants' No. 3, p. 88.

AS TO CHANGE OF PROPORTION.

In machinery, in order to defeat a patent, it is not necessary to show that prior machines existed, in every respect similar to the one patented, as a change of former proportions will not support a patent.

See Woodcock vs. Parker, 1 Gall. Rep. p. 438–440.

Whittemore vs. Cutter, ib., p. 480.

It has even been held, that *putting an old machine* to a *new use*, when the patentee has done this, in connection with the *original discovery of some new quality of matter*, so that he makes a new and useful result, is not patentable.

See *Tatham et al.* vs. *LeRoy*, in the last volume of Howard.

"A fortiori," the lengthening out of the body of the Chapman or Tredgold car, to be used for precisely the same purposes on the same description of railroad, is not patentable: for that amounts only to putting an old machine (with a slight change of proportions) to an old use.

How can it be, that we can use a certain machine of a certain length, with trucks under it, suited to the road, and yet that we cannot use the same machine, made longer in the body, with the same trucks under it?

The only difference, except in the springs and wagon bolsters, between Winans's and prior inventions, is, *a change of proportions*. [Mr. Whiting here illustrated this proposition by explaining the models to the Court; substituting large wheels and long body on the Chapman and Tredgold models, produced the near coupling of the wheels in each truck, and the remote coupling of the trucks.] I refer to your Honor's opinion in the case of Wilbur vs. Beecher, on this point. (Pamphlet Report, p. 14.)

If these principles of law be correct, they answer most, if not all the Plaintiff's criticisms on prior inventions.

If any car like the Quincy car was used only for freight, it is not to be patented by Winans, because he uses it to carry passengers. Especially as his car is a freight as well as a passenger car; and his drawing is of a freight car only.

Whether the car is used to run fast or slow makes no difference.

Whether it is to be drawn one way or the other, is immaterial, since, in fact, no particular mode of traction is claimed.

Allen's carriage was used under a locomotive, yet if it were applied to a passenger car, it would be the same in principle.

If Winans can monopolize all eight-wheel running gear, Allen could not have applied it to his locomotive without infringing.

It is immaterial what load is borne by the platform, whether wood, stone, passengers, or a steam-boiler, or whether the machine move fast or slow, if it only has the capacity for slow and fast motion.

Even if the eight-wheel double-truck carriage had been made to travel on common roads, and had never been applied to railroads, and if Winans had been the first to make that application, he could not sustain a patent for thus applying an old machine to a new use, according to the decided case before cited, viz. Winans vs. The Boston and Providence Railroad Company.

How much less could he sustain a patent for merely making use of the same running gear, because he placed it under one kind of a platform or another, or carried one kind of burden rather than another, or merely altered the proportion of parts, without introducing any new principle of operation.

So also, if the Chapman or Tredgold car, were originally designed merely to distribute the stress of the load, or what is the same thing, the weight of the load—equally on eight wheels—or even if their plan was to place and retain the wheels at equal distances apart, along the rail, would it be any invention to put on a longer body instead of a shorter one, upon the *original* trucks, or to put passengers thereon, instead of freight?

If Tredgold or Chapman may answer a better purpose by merely *prolonging* the body, can there be invention in lengthening that body?

I have already denied such an hypothesis.

THE QUINCY CAR.

The *Quincy car* is the next subject for consideration. The Court has examined the model. It consists of a long bearing platform, which is made of solid timbers fastened together by two cross-pieces at the ends thereof, which, while they unite these long timbers, at the same time constitute the bolster pieces. The bolster pieces are penetrated by king-bolts, which pass through these, and through the middle of the ends of the centre timber; the under part of each bolster is rounded up, and it is also armed with a transom plate and side bearings, which correspond to similar transom plate and side bearings upon the trucks underneath.

There are two trucks, one at each end of the bearing platform, swivelling under it, upon the king-bolts.

Each truck has four wheels, and a solid, rigid rectangular wheel-frame covered by a solid platform, and said wheel-frames have side pieces and double cross bolsters. The axletrees on which the wheels revolve are metal, and bedded in the cross timbers at each end of the wheel-frame. The bearing points of the wheels on either side of the truck, are about the same distance from each other as the width or gauge of the track. The diameter of the wheels is smaller than those in general use at the present time. The trucks are coupled by the body sufficiently remotely from each other to allow each to swivel entirely around without interfering one with the other.

In regard to its mode of operation, it is precisely the same in principle as the eight-wheel cars now in common use; the bearing points of the wheels being equi-distant with the gauge of the track, and the two trucks placed at or near the ends of the bearing platform adapt this car to a distribution of the weight equally upon the wheels, while the swivelling of the truck adapts it to pass smoothly and safely over the straight parts, curves and other inequalities of the road, and the peculiar shape of the bolster adapts it to conform to great and sudden changes of grade, and the side bearings to prevent the body from swaying or tipping one way or the other.

The said Quincy car does not contain the peculiar mode of

uniting the axles of the wheels by a spring bolted to the boxes, and, owing to the small size of the wheels the flanges are not brought as near as possible together without coming in contact; but all that is material and essential in the arrangement of the eight wheels of the car, and the connection thereof with the body, is there reduced to practice in a manner which obtains all the advantages, while it avoids the defects of the arrangement as shown in Winans's specification. The trucks are placed as far apart as the length of the body will permit, and the wheels are brought so near together that their bearing points are as far apart as the width of the track; and this arrangement, on the whole, is better than a closer proximity of the wheels.

Four-wheel railroad cars having been before constructed and in use, and the Quincy car showing the manner in which two trucks could be placed under the ends of a car body, and there being nothing to prevent the builder from making the body as long as the amount of travel or transportation required, or increasing the diameter of the wheels, the constructor would be called on to do nothing else but merely change proportions and to substitute wheels, turning with axles, which were well-known equivalents for wheels turning on axletrees. It is true that various improvements have been applied to modern cars in addition to what is in the Quincy car; but the arrangement of the wheels and construction and connection of the truck with the body of the car, still remains the same in its essential character. And no invention would be requisite, so far as regards the arrangement of the wheels and the connection of the trucks with the body of the cars, to swivel to the curves and run smoothly and safely on the road.

The position of the bearing trucks is at the end of the body, precisely as Winans recommends. See his patent, Plaintiff's proofs, p. 114, fol. 13.

The existence and structure and date of the Quincy car, are proved by

1. Gridley Bryant,   Defendants' No. 3, p. 141.
2. Jotham Cummings,   "   " 2, " 143.
3. Noah Cummings,   "   " 3, " 145.
4. David R. Nash,   "   " 3, " 147.

(5.) John L. Cofran, on the part of the Plaintiff, confirms the same facts.

In regard to the experts on this subject, I will merely make reference to their names and to the pages where their testimony is to be found.

   5. John Murphy,   Defendants' No. 4, p. 7, fol. 27–33.
   6. William Pettit,   "   " 4, " 13, " 51–54.
   7. Septimus Norris,   "   " 4, " 1, " 1–10.
   8. Joseph L. Kite,   "   " 4, " 11, " 41–50.
   9. M. W. Baldwin,   "   " 4, " 3, " 11–16.
 10. Oliver Byrne,   "   " 4, " 9, " 34–40.
 11. Richard French,   "   " 4, " 6, " 21–26.
 12. John L. Cofran, Plaintiff's proofs, p. 74, fol. 349–352.

We have not a greater number of witnesses as to the origin and history of the Quincy car, because it was not supposed to be necessary, the dispute being about the principles of its operation, and not in regard to structure.

It is admitted that the invention and the public use of the Quincy car was prior to Plaintiff's alleged invention. On this ground, therefore, we rest in safety.

The construction of the car, as stated, is admitted. The only questions are, whether it embodies the peculiarities of Winans, or the true principle of the eight-wheel car, or substantially the *principle* of the eight-wheel car *as made by the Defendants?*

The Quincy cars being double truck eight-wheel cars for freight, with swivelling trucks, there is no denying that they embrace and embody all the general mechanical principles and arrangements which are essential to the organization of an eight-wheel double swivelling truck car, either for passengers or for freight.

After reading the affidavits, and particularly that of the distinguished mechanic who has attained so high a reputation in Massachusetts as a Civil Engineer, and inventor of the eight-wheel car, as well as a great variety of other railroad machinery, a man of high character as well as of inventive genius, whose name will be forever preserved in our history as the engineer and builder of the first railroad in the United States,—*Gridley Bryant*,—no one can entertain a doubt that his eight-wheel car, which was in full and successful operation in 1829, was designed and intended to answer the *same purposes*, (as to the transportation of freight,) as Winans's car was designed to answer when used for freight,—to carry the same more smoothly and evenly over the curves, straight parts, inclined planes and inequalities of the road. That the body or platform was made longer or shorter, according to the strength of the materials and the weight they were intended to bear.

The ease of motion attained by this car was produced by the same process, which Winans subsequently adopted, so far as regards the position of the trucks under the EXTREME ENDS of the body or platform, the length or shortness of the body being (by Winans's specification as already stated) immaterial.

So that the language of Winans's specification is an *exact description* of the Quincy car, as to the connection of the truck with the body by the swivelling pin, (excepting that the Quincy car had the addition of side bearings, and so do the Defendants' cars,) and an exact description of the arrangement of wheels in each truck, (except that the Quincy car uses the solid, rigid rectangular wheel-frame, and so do the Defendants, instead of the connection of the axles by the long springs;) and in the Quincy trucks the bearing points of the wheels on the rails are equi-distant with the gauge of the track, (and so are Defendants',) while Winans's are far closer together.

The Winans truck is constructed upon the theory of bringing the axles as nearly as possible to coincide with the radii of the curves of the road, while the Quincy truck and those of the Defendants avoid that theory as pernicious in practice.

Taking the Quincy car, just as it was in 1829, and as it is at the present day, and substitute for the freight platform a car body, and you have all that is really essential in the Defendants' cars.

It is true that many modern improvements have been added to make them pleasant and convenient; but so far as relates to all that Winans claims, viz.: the *arrangement* of the eight-wheels in two swivelling trucks, placed under the ends of the body, and the *connection of the body and trucks* by a king-bolt, so as to allow the trucks to swivel and conform to the curves and inequalities and straight parts and grades and inclined planes of the railroads, these cars are substantially the same.

And if, (as has been so often *asserted* and so often admitted on this trial,) the putting the wheels in each truck a few inches nearer to or more distant from each other, and placing the trucks nearer or more remote, under the body, (especially when they were originally placed under the extreme ends thereof,) cannot be a subject of *a patent*, cannot require invention, being merely an alteration of proportions, and not introducing any new mechanical principle into the structure ; and as no other alteration whatever from the Quincy car is wanted in order to make the Defendants' cars, except merely lengthening the body, it follows that the Defendants do not infringe the Plaintiff's claims by using whatever is embodied in the Quincy car, including the arrangement of wheels there found, their connection with each other, and their connection with the body, and the placing of the trucks at the extreme ends of the body. If they do infringe Plaintiff's claims by such uses, then the Plaintiff claims what was in the Quincy car, and that being a prior invention, the patent is void. If, on the other hand, the Plaintiff's claims do not *cover* those arrangements, but only the peculiar devices of long springs and wagon bolsters, the Defendants do not use them, and, therefore, do not infringe.

I now propose to examine all the objections to the Quincy car that have been made, or that can be anticipated.

1st Objection. It is a car for freight.

Answer. So is Winans's ; and it is immaterial to the construction of the running gear, whether the body or platform carries passengers or freight.

2d. The wheels are smaller than car wheels now used.

Answer. Mere change of proportion is not material, and the patent describes no particular size of wheels.

3d. The flanges of the wheels are not close together in each truck.

Answer. Merely increasing the size of the wheels (i. e. a change of proportions) remedies that matter of the approximation of the flanges ; and putting on larger or smaller wheels does not vary the distance of the axes, nor, consequently, the distance of the *bearing points*, and there is no essential difference, whether the flanges are nearer or further apart.

The Court will understand me as speaking of the wheels in each truck.

The essential distinction is, that the bearing points in the Quincy car are distant from each other as far as the width of the track. It is the same in the Defendants' cars. It is not so in Winans's patent.

4th. That the body, or platform, is not so long as the Winans car, and does not extend the whole length over the platform.

Answer. The length of the body is not material, in principle, nor prescribed in the patent, so long as the bearing of the body or platform on the trucks is at or near the ends.

Whether the car be long or short, the principle is the same.

The platform or frame of the sixteen-wheel car, at Quincy, could have been as well put on to the eight-wheel car, and then the length of the car body would have exceeded Winans's. And what is more important is, that the shortness, or length of the body, is not a material part of Winans's plan.

The Quincy car is as long as Winans requires. See fol. 20, of patent, where he expressly says that the length of the body is not material, if only the trucks are placed at the extreme ends. Winans's patent is, therefore, no more for a long body than a short one. It is expressly applicable to any body or frame that will allow two of his narrow trucks to swivel under or even beyond its ends.

5th. That it was only made to carry heavy stones, that could not be subdivided.

Answer. As I have already stated, it is not material what kind of freight was carried.

6th. That it was not made for rapid motion.

Answer. Neither is it material how fast the owners chose to allow it to run, if it were *capable* of running fast. The principle is applied to and developed and embodied in the organization, and not the use. It would be strange if an ordinary road carriage, usually driven at a slow rate, could be patented because a fast horse were tackled into it.

7th. That the wheels were not of the kind most usually employed on railroads at the present day.

Answer. This is not true in fact. The wheels were constructed with a conical flange, and they operated just as ours now do.

Moreover, CONE wheels had been in use prior to 1825, and they were used in the Quincy cars. See Tredgold, Plate 111, Fig. 24, pp. 42—3; also, George S. Griggs, Defendants' No. 3, p. 104, fol. 414—18, and p. 110, fol. 438—39. And Winans does not designate one kind of wheels more than another as a necessary part of his organization; he claims the arrangement of every description of wheels.

8th. That the wheels turned ON their axles and not with their axes, and that this was less safe.

Answer. (*a.*) This is a matter of practice, not theory.

(*b.*) The Plaintiff's expert, Hibbard, does not assert, neither do any of the Plaintiff's witnesses say that this proposition is true; nor, if true, that it is material.

(*c.*) That the mode of substituting wheels revolving with their axes instead of on the axles, was well known to mechanics at that time, and has been described and shown to this Court, in several treatises published about 1825. As for instance in Tredgold, p. 13; in Wood, p. 77—8; and in Strickland, Plate 51.

(*d.*) It is admitted that it required no invention to make a substitute of these well-known equivalents.

(*e.*) No witness pretends that the slightest difference, practically, is introduced by the substitute.

(*f.*) Winans does not indicate the preference for one over another mode, even in the truck that he specially recommends.

(*g.*) Both kinds were in use at that time, and each had its advantages, therefore this objection is hypercritical.

(*h.*) The only expert of the Plaintiff metaphysical enough to cavil at the Quincy car, is Hibbard. Some of his objections have been stated by the Plaintiff's counsel and answered by us, and we shall notice and answer the remainder.

1st. He assumes that the object of the Quincy, and other cars, was merely the distribution of the weight over a great extent of RAIL.

If so, it would be immaterial. If Bryant, in accomplishing that object, made a perfect eight-wheel car, embracing all the essential principles of that used by Defendants, it is of no importance what his original design was.

But the assumption of Hibbard is incorrect. The real objects for which the car was made and used, are specially stated by the inventor in this affidavit, viz:—" To carry a large load on eight wheels without injury to the road ; to turn the curves freely, descend the inclined plane, and run on the road, carrying the stone smoothly and safely as possible." Df'dts' No. 3, p. 141, fo. 563–4.

2d. Mr. Hibbard's second objection is, that in the bearing carriages the wheels are not so near as in Winans's trucks.

Answer. They are as near as the Defendants place them, and if not, the size of the wheels may be enlarged, and then they will be as near as Winans's.

3d. The third objection is, that the bearing carriages are not so remote from each other as Winans's.

Answer. They are at the extreme ends, which is the plan recommended in the patent.

(*i.*) The chief objection to the Quincy car is that it is was usually drawn by the truck and not by the frame or body.

Answers. 1st. That where there is but one car, and the cars are not in trains, drawing by the truck is as well as drawing by the body, because the unrestrained swivelling of the truck is not useful. It must be controlled when the axes of the wheels are very near together.

2d. Can it require invention to insert the drawing ring in one place rather than the other, after it had been demonstrated that the hind truck could be allowed to swivel freely ?

3d. There were two ways of using the Quincy car ; one by traction and the other by gravity.

On the inclined planes nothing controlled the free swivelling of both trucks. They would be as free to swivel as though the car were pushed by a locomotive and by the body along a level track. The language of the patentee shows, that all the advantages of his arrangement may be obtained by drawing the forward truck by the perch, and he intended that his own car should be so drawn. The other objections of Mr. Hibbard have been answered before.

(*j.*) Mr. Keller adds an objection that the Quincy car had no conical wheels.

1st. This, if true, is of no consequence.

2d. It had conical wheels.

(*k.*) Again Mr. Keller objects, that for want of springs, the Quincy car did not so well adapt itself, to the change of grade.

To this I answer: 1st. That the bolster was rounded up.

2d. It did conform to very steep grades, even to an inclined plane.

These are all the objections.

Only one expert in the country dares *deny*, that the Quincy car contains an embodiment of all the substantial parts of the eight-wheel car of the Defendants.

The Plaintiff and his counsel knew all about this car, as early as June 4, 1853, and therefore can make no excuse that he has no opportunity or time to produce experts.

This fact is significant. The Quincy car has been in successful use from 1829 up to the present time, and had that road required passenger cars, these could have been used at once for that purpose.

The theoretical opinions of Hibbard, who never had anything to do with railroad machinery, and who has no practical knowledge on the subject, and the fine-spun speculations of the adroit and ingenious junior counsel of the Complainant, are all that is offered to over-balance the ponderous weight of authority of scientific and practical men, as to the principles of the Quincy car, and as to its embodiment of all the essential elements of the cars as made by the Defendants.

Now apply the same reasoning, as to the separating of the trucks, to this car as to Tredgold. The lengthening of the body still preserves the equal weight borne by each wheel, and the weight borne by each span of rail.

Also; the objects of the construction of the Quincy car are not left for inference, but are proved and not denied to be, not only to distribute the weight, but also to swivel and run smoothly and evenly.

Can it be while Winans says that the *length* of the body is not *material*, if supported at the two ends, as was the Quincy car, that we are to be deprived of using the Quincy trucks, under the *short or* LONG platforms?

THE ALLEN CAR.

I now pass to the subject of the running gear of Horatio Allen, as applied by him to his steam carriage.

We assert that this invention was prior in time to that of Winans.

Supposing Winans to have organized the running parts of the Columbus as the Plaintiff claims, the earliest date to which their evidence carries the first sketch, which preceded their first experiment is in February or March, 1831. And Mr. Elgar, being contradicted by all the other Plaintiff's witnesses, as well as by the Defendants', and swearing now what he swore he did not know at the Troy case on this very point, March, 1831, is the first moment at which the Plaintiff's "experiments" began.

Horatio Allen's invention was matured in his mind, in 1830. Had he stopped there this would be of no avail, but in the fall and winter of 1830 he made complete working drawings, very little like those that Winans *says* he made in the following March.

We have produced the identical drawings and plan.

The invention was pressed forward to a successful issue. The first machine was finished and put to use in the beginning of the year 1832, and so well was it liked that three others were also built and placed upon the road before the close of the year 1833.

Allen's first machine would have been built before the Columbus, had it not taken so much longer to build a locomtive than an ordinary car. We do not find here any series of failed experiments for three or four years; for the invention was complete so far as relates to the running gear.

To Horatio Allen belongs the honor of having made the first double truck running gear for locomotives in this country, which was not only perfect in respect to the arrangement and connection of the wheels, but also in the application of pedestals to the springs, the same as are now used on all cars and railroads; a fact that may well cause his name to be remembered with honor in connection with the history of railroads in this country.

If Winans failed in the Columbus, as the Plaintiff says, then Allen's invention was completed, while Winans was only making unsatisfactory experiments.

Mr. Keller argues that an invention is nothing in the eye of the law until it is embodied in some form of machinery—it is nothing while it is the subject of mere experiment. And so Allen's invention was nothing until his invention was completed; that did not occur until after July 4, 1831, and therefore Allen cannot be considered prior in time to the Columbus, which is claimed as Winans's invention!

This is his argument to show the alleged priority of Winans.

But in the next breath, it became necessary (in order to avoid the stress of our argument that the invention was *abandoned* from July 3, 1831 to October 1, 1834,) to take a new ground. He then says that Winans had made no complete invention in 1831, but was only trying unsuccessful experiments! and that he did not complete and embody his invention *until* 1834! In the latter supposition, he was three years subsequent to Allen! How unfair to assume such *inconsistent* grounds of argument!

Not only did Allen put his invention into successful operation, while it is said that Winans was experimenting, but he also made a report on the subject of running gear—the first on the subject—in 1832; which was printed, published, and circulated, explaining fully the whole philosophy of its structure, and which Winans some years afterward plagiarized.

I will now give your Honor a description of the model and drawings of the Allen Steam Carriage. It is borne by two trucks; each truck has four wheels in a rectangular rigid wheel frame, which preserves the parallelism of the axes; the points at which the wheels bear upon the rails are separated about equally to the width or gauge of the track, which distance is the most beneficial in actual use. The truck frame is united to the axes of the wheels by means of springs and pedestals similar to those now in general use, which, while it gave ease of motion to the burthen carried, effectually prevented the axes from, at any time, losing their parallelism, by confining the motion allowed by these springs at all times to planes perpendicular to the track and equi-distant from each other; thus the wheels were always kept square on the track. The four and hind wheels of the truck were of different diameters, but this fact is wholly immaterial. Each truck had a bolster running across the centre of the same, from side to side, and this bolster was connected with an upper bolster, on which the steam carriage rested, by means of a large swivelling pivot or king-bolt, operating also as a transom plate, and the trucks swivelling readily and freely to the curves and other inequalities of the road. There were also anti-friction side bearings upon each truck to keep the body of the steam carriage from rocking, and assist in supporting the same. The two trucks were placed so near the ends of the steam carriage, that the ends of the truck frames projected beyond the body; and this position was best calculated to sustain the weight of the body. A part of the body hung down between the two trucks. The body of the steam carriage was long, so that it readily rested on two four-wheel trucks, allowing them to swivel to the curves without interfering with each other; and the distance of the bearing bolsters was nearer to the ends of the body than the position now usually adopted in passenger cars, is from the end of the body platform. The said steam apparatus may be taken off, leaving the bolsters and all other parts as they were, and a platform or body for passengers substituted, without invention, and this carriage will then, as it did before, combine all the mechanical elements of the eight-wheel railroad passenger car, as ordinarily used, embodied in a manner exceedingly well adapted to pass smoothly, steadily and safely over the straight track as well as the curves and irregularities of railroads. Indeed, it contains all the most essential features of the running gear now in general use, and is far better calculated to attain the objects described in said Winans's specification, than the mode of arrangment which is recommended in the patent itself. The whole of the objects or beneficial results set out in said Winans's patent, and much more, are attained in said Allen's Steam Carriage.

The witnesses who prove the Allen engine are—

Horatio Allen, Defendants' No. 3, pp. 60–62, 63, 64, fol. 240 to 254.

Christian E. Detmold, Defendants' No. 3, pp. 56 to 60, fol. 225, 227, 228, p. 58, fol. 232 to 239.

To show that the running gear is identically the same in principle and mode of operation, as that of the cars built by the Defendants, we refer to the testimony of

Charles B. Stuart,	pages 91–2,	fol. 364, 5, 6.
Wm. J. McAlpine,	" 95	" 378, 9, 380.
Vincent Blackburn,	" 98	" 391, 2.
Walter McQueen,	" 102	" 407, 8, 9.
Albert S. Adams,	" 116	" 463, 4, 5.
James H. Anderson,	" 119	" 476, 7.
John B. Winslow,	" 122	" 487, 8.
John Crombie,	" 127, 8	" 508, 9, 10.
H. W. Farley,	" 131	" 521, 22, 23.
Godfrey B. King,	" 8	" 306, 7.
William C. Young,	" 151	" 604, 5, 6.

What are the Plaintiff's criticisms on the running gear of the Allen carriage?

1st. That the running gear was applied to a locomotive, and not to a passenger car. My answers are,

1st. Take off the boiler, and you can substitute a platform for freight or passengers.

2d. This change removes all machinery for locomotion, which is the only alleged impediment.

3d. Such a change requires no invention.

4th. Winans's running gear is as applicable to locomotives, as to freight and passenger cars.

5th. Changing the use is no change of principle.

6th. Cars are now made, called steam cars, having the locomotive in the forward end of the car.

The second criticism is—That the carriage is shorter than the length of 20 to 24 feet, recommended by Winans.

Answer. This is a mere change of proportion, and the length of the body is stated in the specification (fol. 20) to be immaterial.

3d. That the wheels are of different sizes.

Answer. As I have already shown, this is immaterial.

4th. The flues of the engine clogged, and the engine was given up.

Answer. That the running gear was perfect, and was repeated on three other engines. There has been a great improvement made on engines, but not in the running gear; and the running gear was on the same principle as that now in use in all parts of the country.

5th. That the Defendants have failed to state in what manner it was connected to a train of cars. They then *infer* that such connection was fatal to the free swivelling of the hind truck.

Answer. Allen's Report shows that he was fully aware of the importance of sufficient freedom of swivelling. This is therefore an unjust inference.

The same objection may be made to the Plaintiff's patent, for it does not describe or show how it was to be connected with the train; though we have shown by reference to facts *dehors* the patent, that it was to be drawn by the perch.

6th. That the eight wheels were equi-distant, and the two tracks not separated far apart.

Answer. They are the same distance apart from centre to centre in each truck, as those made by the Defendants; and that the distance of the trucks depends wholly on the length of the body, which is a mere change of proportion.

7th. That two of the wheels on each truck were used as driving wheels.

This is not material, unless this fact prevented the truck from swivelling sufficiently freely to enable it to conform to the curves. And the evidence is that the trucks did so swivel; and experiment is better than theory.

Moreover. Absolute freedom of swivelling is not requisite, otherwise the truck might turn crosswise on the track.

The use of the wheels as driving wheels, could make no difference in the swivelling, because the whole truck was free to yield to the guidance of the rails, and it would make no practical difference whether the tractive power was communicated through the king-bolt, or directly communicated to the axles.

8th. That the connection of the steam machinery with the running gear, is not shown on the model.

Answer 1st. It was not necessary.

2d. It is shown in the drawing.

3d. It is stated in Allen's affidavit, and explained to have been such as to allow of perfect freedom of trucks to swivel. This is an unfair objection.

Thus we find in Allen all that is essential in the Plaintiff's cars, and all he claims, except the *spring trucks* and the *wagon bolsters*.

1st. The two four-wheel trucks.

2d. Their arrangement at or near the ends of the body.
3d. The wheels and flanges very close.
4th. The trucks as far apart as conveniently may be.
5th. The trucks free to swivel.
6th. The purpose, viz., the distribution of weight, and ease and safety of motion.

The Plaintiff's Counsel has said, that if these cars would suggest the idea of the eight-wheel car, Allen was just the man to make one; and so we say he would have done, if such cars had been wanted, for it required no invention to do so.

Merely taking off the boiler, and putting on a long body, would remove every objection of the Plaintiff.

This would merely be applying Allen's invention to a new use, making only an obvious change.

But the railroad companies were not in favor of eight-wheel cars at that time, nor for many years after Winans's patent was dated; and therefore Allen had no occasion to apply his invention to that use.

This carriage, then, (as well as the Quincy car,) is conclusive, taken in connection with the drawings and the reports, which states:

1st. The objects to be attained.
2d. The mode of attaining them.
3d. Their success.

Who denies that the Allen carriage embodies all the principles of Winans's claim? if he claims any thing beyond the above mentioned peculiar devices. And if any—on what ground?

Not one expert has denied it; and no one but the Counsel asserts it. Can the Court hesitate in their decision?

THE BALTIMORE CARS.

I now pass, with great brevity, to say a few words on the Baltimore Timber, Trussell, Wood and Platform cars. They were constructed as follows:—

Two ordinary bearing carriages, or trucks, (such as were in common use at that time on the Baltimore and other roads,) having rigid rectangular wheel-frames, bearing four wheels each, and having the flanges distant from six inches to two feet, were connected together by a long platform composed of string-pieces of timber; these string-pieces were bolted on top of an upper bolster, so as to constitute a frame-work platform; the bolsters of this platform rested at each end upon two corresponding *under bolsters*, which last mentioned bolsters were fastened upon the bearing carriage. A king-bolt or centre pin connected the upper and lower bolsters, so that the trucks were free to swivel to the curves of the road. The bolsters were under the ends of the platform, and the trucks were drawn by the perch. The wheels in each truck were close together, in some of the trucks, and farther apart in others; but in all cases the two trucks were placed as far apart from each other as the length of the long string-pieces would allow. So that this rude structure combined all the essential elements of the common double truck eight-wheel swivelling car. When used for carrying cord-wood, upright standards were added to the platform, in order to keep the wood in place. When horses and carriages were to be transported, cars were built with long *trussells* instead of a mere platform.

It is proved beyond a doubt,

1st. That the "Timber cars" were running on the Baltimore and Ohio Railroad in May, 1830.

2d. That the "Wood cars" were running in November, 1830.

3d. That the "Trussell cars" were running in December, 1830, being at least six months before the Columbus was finished, and four months before she was begun, and before the plan was produced by Winans. We have shown, conclusively, that Mr. Elgar was mistaken in HIS dates, and that his evidence has been overthrown by all the other testimony in the case, on the part of the Plaintiff as well as the Defendants.

And I take pleasure in saying, that I do not believe that any one of the witnesses, whose testimony is used on either side, intended to make false statements; but that any witness may have innocently made mistakes of facts and dates, is most obvious.

The witnesses for the Plaintiff, who in any wise conflict with our testimony on these cars, are the following, viz. :—

1. John Elgar.
2. William Woodville.
3. Michael M. Glenn.
4. Oliver Cromwell.
5. Thomas Walmsley.
6. Henry R. Reynolds.
7. John P. Mittan.
8. B. H. Latrobe.

Those who must know, and who do not dispute the facts as we state them, as to the "Trussell car," are:—

1. Philip E. Thomas, President of Baltimore and Ohio Railroad Company.
2. George Brown, Treasurer.
3. Lloyd Claridge, Carriage painter.
4. John Ferry, in the Transportation Department.
5. Thomas Davis, Conductor.
6. John P. Mittan, Carpenter.

Your Honor will see that pretty nearly half the witnesses connected with the Baltimore and Ohio Railroad do not deny the fact which we assert, viz.: that these Wood and Trussell cars existed at the time when we state they were on the road. And if you will notice the extraordinary enterprise in procuring, and the adroitness in framing the affidavits on the part of the Plaintiff, such a fact would not have been overlooked; it would have been positively denied unless it were true.

This argument is strong. Four only out of the eight witnesses produced by the Plaintiff, swear positively that these cars did not exist.

John Elgar states his belief;—William Woodville is positive;—Michael M. Glenn rather dodges the question as to the "Wood cars," and contradicts himself as to the "Trussell cars;"—Oliver Cromwell "verily believes;"—Henry R. Reynolds is positive;—Thomas Walmsley has an impression, but is not positive whether the Wood cars were used before 1830 or not.

But all the witnesses throughout the case, and the counsel on the other side, have admitted the preëxistence of the Timber cars.

Now it will be one of the pleasantest duties of your Honor to reconcile the testimony on both sides. This can easily be done.

The counsel on the other side admit that double truck Timber cars existed before the Columbus, though they say that they were not fastened together, so as to constitute an organized machine. Our witnesses say that the machines were organized at that time. Nothing further was necessary than to drive four spikes into the string pieces, and that would be all that was to be done, in order to complete their organization. And the only question on which the dispute arises, is, that whether those spikes were driven so that the Timber cars had their string pieces fastened to the bolsters, so as to make a platform; or whether they were left to slip about upon the bolsters, as chance or accident might require.

Now as the load of timber upon these stringers and bolsters, would, by its own weight, keep the mass from slipping or swaying on the bolsters, no one, except his attention were particularly called to it, would notice whether the two string pieces were bolted to the bolsters or not. And when a change was made from the first rude manner of laying the timber across the two bolsters of the two trucks, to the more convenient mode of fastening the outside stringers to the bolster, is it not possible that some of the witnesses might not have noticed so slight a change, and especially as it is certain that their attention was not drawn to it?

It is another fact, which all admit, that the double truck "Wood and Trussell cars" existed at *some* time. The difference between the witnesses is only a question of *date*. And I will call your Honor's attention to the fact, that one of the Plaintiff's witnesses, upon whom they rely, states that he thinks he saw them in 1832. If some of Plaintiff's witnesses did not see them until 1835, and others saw them in 1832, then it follows that the cars were in existence in 1832; and if they were in existence three years before they were seen by these persons, it is quite likely that they may have been there five years without their knowing it. This, we say, was the fact.

The Plaintiff's witnesses were not the persons who made the cars, while the Defendants' witnesses originally made and subsequently repaired them. A number of the witnesses for the Defendants actually left the road and left that part of the country; these cars being at the time in daily use, long before the date at which the Plaintiff's witnesses say that they made their *first* appearance on the road, so that there is no possibility of a mistake about the fact. Those who say that they did not see them, we presume, did not notice them; nevertheless the cars were there, else they could not have been seen by so many others.

Some of the Plaintiff's witnesses admit what we say, in part.

May, the Plaintiff's witness, p. 124, fol. 542, says that he thinks he saw them in 1832.

If so, all the Plaintiff's other witnesses are wrong. If one of them did see them in 1832, then they were there, and all the evidence of those who did not see them goes for nothing.

The witnesses on the part of the Defendants, who fix the time when these cars were built and put in use, are:—

1. Jacob Rupp.
2. Leonard Forest.
3. John Rupp.
4. Conduce Gatch.
5. Edward Gillingham.
6. William E. Rutter.
7. John H. McClain.
8. Edward May.
9. Henry Shultz.
10. John M. Eichelbergher.
11. James B. Dorsey.

When eleven men swear that they actually saw these cars in existence, at a certain time, will the Court say that the testimony of six negative witnesses shall outweigh them on a point of date?

The existence of the "Trussell cars," as early as 1830, is proved by documentary evidence, in the pay-roll of December, 1830, made by Conduce Gatch, which also mentions that they were made in that month, for the transportation of horses and carriages.

Additional proof is found by reference to the Baltimore Gazette of Dec. 17th, 1830, and the Baltimore American of Dec. 18th, 1830, which mention the fact that the wife of Ex-President

Adams was at that time transported on the railroad, and that her horses and carriage were placed in a *Trussell car*, attached to the train; and the Baltimore Gazette of Jan. 19th, 1831, Defendants' No. 3, p. 46, mentions the Wood cars.

It is, therefore, indisputably proved that these double truck eight-wheel cars existed and were in public use on the Baltimore and Ohio Road prior to the Columbus.

Next in order came the " Columbus," and we hold that it is indisputably *proved* that the person who planned the running gear of the Columbus, and its connection with the body,—being in fact nothing new,—was *Conduce Gatch*, and not Ross Winans.

Who arranged the " RUNNING GEAR " of the Columbus ?

We admit, that the size and shape of the *body*, although merely copied from Smith's model, which was placed in the office of the Baltimore and Ohio Railroad Company as early as 1830, was then first introduced on to the Baltimore and Ohio road; and that this car body, by its unusual length, being twice as long as the four-wheel cars for passengers, would attract attention, even though the new body was in fact placed on a *platform* constructed in the old way, with *the old* running gear.

The testimony on the subject, as to who made the Columbus, is conflicting, if you do not discriminate between the making of the car as *a whole* and the *mere arrangement* of the running gear; but it will not be difficult to settle the question, who arranged the latter.

The Plaintiff attempts to show that Winans was the person who first planned the running parts of the Columbus, and to show this, he relies on a certain working drawing, from which *he says* the car was actually built.

He finds a number of witnesses who say, that he "*produced*" this drawing.

No man swears that Winans made it, and what is a little singular, Winans himself does not state in any part of his own affidavit that he made it or directed it to be made.

The next question is, whether that drawing thus produced, contained, at any time before the car was actually built, the running gear ?

On this, the weight of the testimony is decidedly with the Defendants.

Suspicion is thrown upon the drawing itself, which has been produced.

1st. It is smoked to give it the appearance of age.

2d. It does not look like a drawing that had been worked from in a shop. And if your Honor will compare it with the Allen drawing, which is in reality a working drawing, the difference will be striking.

3d. It has been altered in many important particulars, from its original state.

Any tampering with a drawing destroys its force as evidence, except as against the party producing it.

4th. The car Columbus as actually built, did not correspond with the drawing in about seventeen particulars; but I shall not occupy the time of the Court by dwelling on one of them, they have already been pointed out by my colleague.

5th. *The positive testimony of those who* MADE *the running gear is, that there was no running gear on the plan.*

6th. The uselessness of making a drawing of the running gear when there was no alteration to be made from the ordinary running gear, that had been previously used on the Timber, Wood and Trussell cars.

7th. As to the dimensions alleged to have been taken by Conduce Gatch from the plan, it was necessary to measure the body, in order to ascertain the position of the king-bolt, and thereby, ascertain the *length of the perch* suitable to the car. No one says, that he took the slightest admeasurment of the running gear, or any part of it.

Therefore if Winans made the drawing, it did not show the running gear, but only the body.

As to the directions to workmen said to have been given by Winans, the Plaintiff's witnesses state, that he gave some directions, but they do not mention one word that he said.

On the other hand, the Defendants' witnesses swear that all the directions he gave related to other subjects, and were subsequent to the building of the Columbus.

The Columbus was built at the Mount Clare shop, and the directions were given at the Charles Street shop, and at George Gillingham's shop, where Winans's patent *friction wheels* were made.

The mere vague *reputation* relied on by the Plaintiff's counsel, that Winans was the inventor, and was talked of by some individuals as the inventor, is the most flimsy species of testimony.

The reputation among the makers of the trucks, and many others, was wholly in favor of Gatch, as the inventor of the arrangement of the running parts.

Winans might have styled himself the inventor of the car, and he might have been called so by others, but though he only put a different body, on the common running gear, yet it might have been called " Winans's car," in that sense of the word.

Many no doubt thought that the car, as a whole, was new, although there was nothing new in it in reality, except the body; and although the connection of the wheels with each other on trucks, and the connection of the trucks with the body, was the same as in the preceding Wood and Timber cars; yet, applying the old arrangement to a new and large body, and to the new use of transporting passengers, nothing would be more natural than to call it the Winans's car; and to name Winans as the inventor of it, and that reputation is in no degree inconsistent with the testimony of the Defendants' witnesses, which is, that so far as relates to the arrangement of the *running parts,* Conduce Gatch first invented that arrangement and applied it to the cars which preceded the Columbus; and afterwards, when the Columbus was built, applied the same arrangement to it.

If that suggestion shall enable the Court to reconcile a mass of testimony which would otherwise be conflicting, I shall feel myself rewarded for the time I have spent in explaining it.

Now, has your Honor, in the course of your judicial experience, tried a single case in which the Plaintiff has laid claim to an invention, of any description, without proving a solitary fact connected with its early history beyond this; that he had in his possession at a cartain period a certain drawing; and such an one as I have demonstrated this to have been?

As it is therefore not proved that Winans actually made any drawing from which the Columbus was built, upon what evidence does the Plaintiff rely to show that allegation?

Certain witnesses are produced, who attribute the invention, as a matter of *rumor,* to the Plaintiff, and who also say, that they never heard of any one else claiming it.

Not one witness proves a fact, except from mere hearsay, to show that Winans invented or devised any thing.

Not one swears to a single specific direction given by Winans to any person about the running gear.

Not a man who worked on the running gear ever heard any direction given by Winans on the subject.

The Counsel rely on mere hearsay of gentlemen in office, and not on the hearsay of the men who worked on the trucks.

It would have been easy for Winans to have manufactured this hearsay, by conversation with his friends.

These gentlemen might have been easily deceived about the eight-wheel cars, owing to the fact that Winans had gotten up numerous four-wheel cars, known as Winans's cars.

By comparing the position and opportunity for knowledge of these two classes, the Court cannot hesitate to say, that the presumption would be, that those actually engaged in the manufacture of the Columbus, would best know who devised the plan of the different parts, and who gave them directions about the work.

The preponderance of evidence, therefore, is decidedly with the Defendants.

Among other corroborating circumstances, I may mention that all new improvements were specially noticed, and credit given to the supposed inventor, in the " American Railroad Journal."

The Columbus is not noticed as a new invention. The eight-wheel car was occasionally mentioned, though not in connection with Ross Winans as the inventor.

The Engineer's reports to Mr. Thomas, the President of the Baltimore and Ohio Railroad, never mentions Winans as the inventor.

The President's report, in October 6th, is the first publication that mentions the thing in connection with Winans's name.

The peculiar circumstances under which that report was made, show the mention of his name to have been a matter of arrangement between Mr. Thomas and Mr. Winans; Mr. Thomas professing to recapitulate facts stated in the reports made by the Engineer and others to him, and referring to those reports as the foundation on which his statements rest, is not *justified* in any such statement by the reports of Knight and Gillingham, to which *he* specially referred.

Knight was the superintendent of the road, and Gillingham was the master of machinery, and therefore it is strange that they should not have noticed, in their reports, so important a fact as the invention of a large passenger car, upon a new principle, if such an invention had been supposed by them to have been made by Winans.

Ross Winans himself has not stated in his affidavit, that he did devise or have any thing to do with the running gear of the Columbus.

And it seems to me most extraordinary, if he did invent it, that he should rely upon mere hearsay to prove the fact, and upon a drawing which has been tampered with; and that he should not even have asserted the fact upon his *own oath*, while he has stated many other things in his affidavit, which are of far less importance than this would be, if true.

Now the *burden of proof is upon the Plaintiff, to show that he did invent the Columbus*, the alleged invention being so long anterior to his patent. And this is true, notwithstanding the patent is itself prima facie evidence of the originality of the invention there claimed.

Has the *Plaintiff* sustained this burden of proof? So far from it, we assert that the weight of the Defendants' testimony, which is positive, *far* exceeds that of the Plaintiff, which rests upon mere hearsay.

And it follows, that Winans, not being proved to have been the inventor of the running gear of the Columbus, his invention will take date, not in 1831, but in 1834; thus giving priority to many cars which were of a date subsequent to 1831.

JERVIS'S ENGINE.

I just allude, in passing, to the fact, that in 1833, Jervis produced his engine, a printed description of which is in the "American Railroad Journal."

I mention this in its chronological order, and refer to the following witnesses:

    1st. John Edgar Thompson.

2d.   Asa Whitney.
3d.   David Mathews.

[Mr. Whiting here described the running gear of the Jervis locomotive, which had ONE four-wheel swivelling truck under the forward end of the engine, and then proceeded :]

And, Sir, there is nothing left to be done but a duplication of trucks under the same engine, to make the running gear like that of the Defendants. With that remark, I dismiss this machine.

FAIRLAMB'S CAR.

I now pass to the patent granted to Fairlamb, and refer to the testimony of

1st.   Jonas P. Fairlamb.
2d.   Isaac Knight.
3d.   George Beach.

And, if the Plaintiff did not invent the Columbus, and embody his perfected invention therein, Fairlamb will precede Winans.

In the drawings of Fairlamb are clearly shown the peculiarities claimed in said Winans's patent, excepting that the axles of the wheels are borne by a rigid rectangular wheel-frame, and not connected together by yielding springs. The close proximity of the flanges of the wheels, in each truck, is there shown, the flanges being represented as but a very few inches apart. The trucks are constructed in two ways ; one allows the axles of the wheels a certain limited motion in the truck frame itself, with a view of allowing it to conform to sharp curves, as represented in Figure 1 and 2, while the other truck in Figure 2 is constructed in the ordinary manner, allowing no play to the axes ; each of these trucks swivel under the body by means of large transom plates, and are placed near the ends of the body of the car. Fairlamb's drawings embrace all of Winans's arrangement. One of the trucks in Figure 2 allows the axles to play, and the other truck is the rigid wheel-frame holding the axles parallel, as is the case in the cars in general use. So far, therefore, as regards the near coupling of the wheels in each truck, and the remoteness of the trucks from each other, it is identically the same as Mr. Winans's.

A mechanic of ordinary skill in car building, having laid Fairlamb's drawings before him, and adopting the truck in Figure 2, which holds the axes parallel, or in other words omitting the apparatus which permits the axes to vibrate, and constructing both trucks alike with parallel axles, would have nothing to do but copy the drawings in order to construct an eight-wheel car, such as is in common use, excepting that the wheels in each truck would be closer together than those in general use, thereby more resembling the arrangement claimed by said Winans.

Thus the drawings show completely and distinctly a double truck railroad car, (unfortunately) so like that described by Winans in his patent, as to be of little use at the present time.

The drawing is substantially a copy of the same that was attached to the original patent, although both had been restored. See Fairlamb's affidavit.

If Fairlamb had arrived in 1832, as he states, to this arrangement of the wheels and trucks which Winans particularly describes and recommends in 1834, he had only arrived at an erroneous conclusion, and one useless in practice.

We do not say that this patent was prior to the Columbus, but that, if Winans has failed to sustain the burden of proof, to show that he invented the Columbus, then Fairlamb is prior to Winans, the date of the patent being January 19, 1833.

Next came the "Winchester," which was put on the road in March, 1834.

THE VICTORY CAR.

Then came the "Victory," of Philadelphia. The plans of this car were completed in June, 1834, and she was commenced in August, 1834; so that long before his patent, Winans had an opportunity to see every thing that was contained in the car.

The Victory was completed on the 3d of July, 1835; she was thirty-five feet long, and was then put in use, and she was the predecessor of the Washington cars.

But the builders fell into the mistake of putting the wheels too close together, perhaps copying the mistaken philosophy of Fairlamb or somebody else. They had to alter the trucks by spreading the wheels farther apart. See French's affidavit, p. 5, No. 4.

The witnesses as to the " Victory," are,

1st. Laban B. Proctor,   Defendants' No. 3, p. 7.
2d.   John Murphy,   "   "   9.
3d.   William Pettit,   "   "   12.
4th. Joseph L. Kite,   "   "   32.
5th. Harmon Yerkes,   "   "   35.
6th. Jacob C. Carncross,   "   "   36.

Next to the Victory, (which drew by the body,) came

THE WASHINGTON CARS.

These cars were not begun till *November*, 1834, more than a month after the patent was issued, and they were not, at first, planned to draw by the body, but were intended to be drawn by the perch, like the others which had preceded them; and this alteration was suggested by Rupp, after building the freight cars. See testimony of Rupp and others.

I will now mention the order in which these cars succeeded each other.

1812. The Chapman car.
1825. The Tredgold car.
1825. The   "   eight-wheel steam carriage.
1829. The Quincy car.
1830. The Baltimore timber cars.
1830. The Baltimore wood cars,
1830. The Baltimore trussell cars.
1830. The Allen car or carriage.
1831. The Killeyney-Kill and Dalkey car.
1831. The Columbus, and great numbers of wood, and horse and carriage cars.
1833. The John B. Jervis truck engine.
1834. The Fairlamb Patent.
1833–4. The Dromedary, the Winchester, and the Comet.

June, 1834. Drawings made of the Victory in Philadelphia.
Aug. 1834. The model of the Victory was made, and the car commenced.
Oct. 1st, 1834. Winans's Patent.
July 3d, 1835. The car Victory completed, and in use, being the predecessor of the Washington cars.
1835. Rupp's 110 Freight cars.
1835. The Washington cars.
1835. The Alleghany, and numbers of cars at Philadelphia.

Such is a description of the various inventions which preceded the Columbus, and also of those which preceded the Plaintiff's patent, and such are our answers to the criticisms made upon their structure.

STATE OF THE ART.

It will now be my purpose to give a brief summary of the state of the art of building eight-wheel cars up to March, 1831; and before the Columbus was begun, the following parts and combinations were known and in public use.

1st. The car body, at least twice as long as the common four-wheel car, as were the Trussell and Wood cars, &c., and the freight cars on the Killeyney-Kill and Dalkey Railroad. Indeed, using two trucks required the body to be about twice as long as a single truck.

2d. The apparatus for allowing the trucks to swivel under the body, consisting of the king-bolt and transom plate, &c.

3d. The two trucks, one under each end of the body or platform.

4th. The wheels in each truck, put within six inches of each other, while others had the bearing points equi-distant with the gauge of the rails, and at all intermediate distances. See Mc Mechen, p. 41, Defendants' No. 3.

5th. The trucks placed at the ends of the body, as in the Quincy car, and near it as in the Trussell and other cars.

6th. Wheels of different sizes, large or small;—some revolving on the axes and some with the axes.

7th. Springs applied, in various ways, to eight and four-wheel carriages; not, however, like Winans's; but like those now in use, as the Allen carriage, and in the common four-wheel trucks mentioned in the patent.

India rubber was applied by Conduce Gatch under the bolster in the summer of 1831, which was suggested when the Columbus first went out. Tredgold also mentions them, and other works shown to the Court.

8th. The rigid rectangular wheel frame was also in common use, though not adopted by Winans.

9th. Side bearings were also used, though not adopted by Winans, also the trucks and the draft by the middle of the ends of the body.

10th. Thus, so far as relates to the arrangement of the wheels, as to distance from each other, and as to the remoteness of trucks from each other, no change had to be made from what was well known, even in building the Columbus.

The only change, as before stated, was in 1st. The springs; 2d. The wagon bolster.

11th. Nothing was left to be added, even to the philosophy of the subject, and no new principle to be illustrated after Allen's Report in 1831 and Jervis's Report in 1833.

Winans's specification was a mere re-hash of Jervis and Allen.

12th. Even the long body had been extended far longer than Winans describes. The Victory, begun in September, 1834, was thirty-five feet long, which was three times as long as the four-wheel cars.

Here we have all the *principles* of the eight-wheel car, with *a great variety of modifications, embracing all that the Plaintiff claims, except* THE SPRING TRUCKS *and* WAGON BOLSTERS.

The Plaintiff attempts to show that at the time the Columbus was introduced, a new era had arisen in the history of railroads; that greatly increased rates of speed were required, and that, therefore, some new arrangement of the wheels became necessary, to meet the emergency.

This is not true; but the fact is, that the new era consisted in the improvement of the locomotive.

1st. In traction by friction.

2d. By turning the waste steam up the funnel.

The running gear, then already made, was as well adapted to running fast as slow.

The great step from the four-wheel car to the double swivelling trucks, had already been taken before Winans's appeared, by Chapman, Tredgold, Allen, by Bryant, at Quincy, and by the Wood and Trussell cars, at Baltimore.

There was nothing left to do, except to make such a change in the proportions of parts, as experience should determine.

The eight-wheel car, as known and described and used before 1830, was adapted to *any size, any length, any distance of wheels, any curves and any inequalities.*

The demands caused by increased business and speed required greater and increasing departure from what Winans describes as *essential.*

PROGRESS OF INVENTION.

Every invention may be said to have its *own history.* I had intended to attempt to develop the progress of improvement in the eight-wheel car; but time will not allow me to go into details.

The Wood, Trussell, Quincy, Allen and other cars had been the legitimate product of the necessities of railroads (having sharp curvatures) in the earliest periods of their existence.

It is most probable that inventions resulting from the actual and practical necessities of the case, should be made by those who are earliest concerned in building the roads and in constructing the machinery necessary to operate them. As my able colleague has well said, in 1828, or thereabouts, Ross Winans was a common mechanic, having nothing to do whatever with the subject of railroads. Conduce Gatch was a regular practical millwright and machinist, devoted to that branch of business which has educated such men as Oliver Evans, and others of the most distinguished machinists and mechanics and inventors of the day. Mr. Gatch was a shrewd and close observer, a fine mechanic, a practical man, greatly relied upon and respected. He is attacked because HE CLAIMS *some inventions!* The *evidence* shows how his *mind* was at *work.*

Any man would observe, in shoving a railway truck around a curve, or in seeing a horse draw it, that coupling the axles of the wheels within a few feet of each other would make the truck run easier. This was a common, well-known fact. For that reason the wheels were placed near together on the four-wheel cars, and the four-wheel car was in fact so constructed as readily to pass all the curvatures and inequalities of the road. Then it became necessary, in the *construction of the roads*, to transport long timbers, and as they could not be placed, without danger, upon a single truck, they were borne by two bearing carriages; and as it was perceived that in turning curves the bearing carriages would swivel *under the timbers*, therefore, it was necessary to put on a swivelling bolster, which was a common device, to enable the *trucks* to swivel without *displacing the load*.

Then it became necessary to carry long loads of cord wood and other long freight, like carriages and horses, which could not be contained in the length of a four-wheel car.

In order to do this, they constructed a bearing carriage, *fastening the timbers to the bolsters*, so as to make an organized rigid bearing platform. They also made a Trussell body, resting at its ends upon the two trucks, to carry carriages and horses all harnessed together, like that of the wife of Ex-President Adams, on December 17th, 1830.

This running gear embodied a nearness of the wheels in each truck, and a remote coupling of the two trucks, fully equal to that which is used at the present day; and moreover, if the ordinary trucks in use at that time were the ones put under the Trussell cars and the Wood cars, those cars would come under the precise terms of Winans's patent, where he states that the ordinary bearing carriage may be used, unless indeed a variation of some six or seven inches between the axes in each bearing carriage is supposed to introduce an entirely new mechanical organization and mode of operation in the truck.

In fact, the centres of the axes in the fictitious drawing annexed to his patent, show the distance between the axes to be about three feet, while the patent itself admits that the ordinary bearing carriages, or some of them, had their axes only three and a half feet apart. In reality however, the evidence shows that the distance between the axes of the bearing carriages had been much nearer than three and a half feet.

Indeed, no single advancement in the development of the railroad machinery is attributable to Winans, though eager to grasp at every patent of every trifling, supposed improvement.

Not a single one of the pretended improvements of Ross Winans before the date of his patent of October 1, 1834, which were ushered into the world with such pomp and parade, is in use, at the present time.

The pretence that Winans made the first suggestion, as to the application of SPRINGS to trucks, is ABSURD as well as untrue. Such springs were in common use. They were described in all the books; and the pedestal springs now used were actually IN USE on *Allen's locomotive* before Winans ever spoke of them to ANY ONE.

Even the present form of the CAR BODY (change from the old *coach* body to the SQUARE now used) was first made and modelled by JOSEPH SMITH of Philadelphia in 1828 or 1830:
(See affid. of JACOB SHRYACH, p. 39, No. 3, etc.)
which pattern Plaintiff BORROWED in 1837 or 8, to MAKE a DRAWING of.

The MODEL of Smith's CAR was made *in Baltimore*, Plaintiff's residence, and it was left by Smith at the OFFICE of the Baltimore and Ohio Railroad Company the SAME YEAR, and undoubtedly was seen by Winans before he began to think about the Columbus.

Even his pretended improvement in the construction of the Washington cars, which have been alleged to be a sort of pattern for those which have followed, was already invented; and the cars were under way and first put in use at Philadelphia before the Washington cars were built. And the plan of the Philadelphia car is undoubtedly the progenitor of the plan of the Washington cars.

Therefore we cannot trace the hand of Ross Winans, in truth, to any one of these improvements. The fact that Winans was an assistant to a *civil* engineer, even if it were inferred from that fact that it was his duty to invent or suggest improvements, has no tendency to show that he did make any substantial and valuable invention. And while he made certain supposed improvements like the friction-wheel, which though extravagantly praised, turned out worthless. The country is under no obligation to this gentleman.

The only way that Winans is known to the country is, that he and Gillingham purchased the patent of Phineas Davis, for the chilled wheels, with a wrought iron ring in them, after Phineas Davis, in 1834, was accidentally killed by a locomotive on the Baltimore and Ohio Railroad; and by the notorious failure of his Crab engines in Massachusetts.

The only instance in which the Winans' car has been TRIED, was with SADDLES to the springs, and even then it was CONDEMNED as WORTHLESS.

See affid. JACOB SHRYACH, p. 40, No. 3, and others.

Indeed so far from the present eight-wheel cars being indebted to Winans, they have many improvements not thought of in Winans's time; and a different construction and arrangement, and numerous improvements or inventions, not described in Mr. Winans's specification.

They have rigid square-sided wheel frames, pendulum or swinging bolsters; they have male and female transom plates for the trucks to swivel upon; have pedestals and springs to allow the axles to move vertically and still keep them parallel with each other, and the wheels square on the track.

Some of the cars have India rubber springs,

Patent lubricating boxes,

Patent soft metal bearings,

Safety beams in the truck frame, to hold it up if an axle breaks.

Improved brakes,

Changing backs to the seats,

Ventilators,

Draw spring couplings,

Patent car wheels,

And car bodies more than double or threefold the length of the body mentioned in Winans's patent, and about five times the length of the ordinary four-wheel car.

And no single improvement used at the present day owes its origin to the invention of Mr. Winans.

NO NEW PRINCIPLE.

The *validity of Winans's patent* depends upon his having *discovered and embodied*, and clearly claimed, *some new law or principle of mechanics*, as applied to the running gear, or else to some *peculiarity of construction* or arrangement not known before, and which is clearly distinguishable from what was known.

1st. Has he discovered any *new principle or law of mechanics* and *embodied it in some form*?

He has not; and this is evident from various considerations.

1st. The patent itself, which displays the philosophy of his invention, does not set forth any new principle, such as would have been claimed by the first inventor of a double truck swivelling car. For doubtless such a car does embrace and embody mechanical principles of action and construction and operation, taking the car as a whole, which are not embodied in a *four-wheel* car.

On the contrary, the patent admits the pre-existence of all the parts of the car, and of the same *general arrangement* of parts; but claims a peculiar arrangement of the wheels with each other, and a particular connection of the truck with the body.

2d. Many double truck swivelling eight-wheel cars having previously existed, which would answer, although perhaps less perfectly, the same purposes claimed to be answered by Winans's; every office that the said Winans's truck can be claimed by his counsel to perform, may be shown to have been performed by the trucks that preceded him.

To wit—the swivelling of the two trucks to the curves under one body, and the near and distant coupling of the wheels necessary to sustain the body, and to develop the swivelling principle with steadiness and ease of motion, to allow of a high speed.

These were in the Timber cars, the Wood cars, the Trussell cars, the Quincy cars, the Allen Steam Carriage and in the Columbus, in use in the United States, and are in the descriptions of Chapman and of Tredgold.

So far as any peculiar mechanical principle can be alleged to have been developed or embodied in Winans's car, those principles were not only already in use, but the rationale was already also made known to the world in the publications of Allen and Jervis, Chapman and Tredgold; so that for this reason, also, it cannot be pretended that Winans had discovered or made to the world in his specification any new principle or theory of mechanical philosophy, or had shown the application of any old or well-known principles in a new way. The fact that he is not able to make any clear and definitive distinction between what he claims as his, and what he admits was old, is itself evidence that there is nothing new in principle.

There is no language in his patent calculated to indicate the discovery or application of any thing new in principle; it all relates to mere alteration of proportions and arrangement, for the purpose of bettering a result already, to a considerable extent, allowed to have been attained in prior structures.

And although Winans's THEORY of making the axes of the wheels coincide as nearly as possible with the radii of the curves, was merely stating on paper what had been before reduced to practice, yet he may well have the credit of producing that *theory*, which is not itself the discovery of any mechanical principle, but is only an erroneous idea and mistaken application of the old and established principle of the swivelling trucks.

Suppose, for ARGUMENT'S SAKE, that Plaintiff had been the first to DISCOVER AND EMBODY the "SWIVELLING PRINCIPLE," PERHAPS he might have covered all similar PRACTICAL MODES of embodying the same principle.

But there is NO PRETENCE that Plaintiff first discovered this VITAL AND ESSENTIAL principle of the LONG-BODIED CAR.

The IDEA or principle that Plaintiff's counsel ALLEGE he discovered is, that in EIGHT-WHEEL CARS the trucks should be allowed to swivel ABSOLUTELY FREELY, by drawing by the body.

Now this is not a PRINCIPLE that Plaintiff DISCOVERS; it is not a *new idea;* but is a MERE mode of USING an old *invention of the swivelling truck in* ONE *way instead of another*, for the SAME PURPOSE.

It is *no more* the subject of a PATENT than it would be to draw a WAGON by the body instead of the shafts.

As *Judge Nelson* said in *Wilbur vs. Beecher*, p. 14,—" A person operating a combination already discovered, may, by experience in its practical use, see where it can be altered, and may call in a mechanic and have an alteration made which may improve *the machine*. This is a necessary consequence of the practical use of a machine by a man of ordinary skill and judgment, but there is no novelty or invention in such alterations."

As to the alleged NEW IDEA that the *trucks* would swivel more freely when the draft was through THE BODY, this is no new idea, as I have shown it had long before been embodied in various FOUR AND EIGHT-WHEEL CARRIAGES, and the whole philosophy of the subject explained, published and printed long before 1834. AND THAT such claim can form no part of Plaintiff's patent.

Suppose that Plaintiff had been the *first* to discover a theoretical idea that the trucks must be allowed to swivel with absolute freedom, and suppose that he had built a car accordingly and had put it into successful operation and then taken out the patent in the TERMS of Winans's patent, he could not, even then, PREVAIL IN THIS CASE.

1st. Because he has not described or set forth in any part of his specification any such IDEA OR PRINCIPLE.

2d. He HAS NOT claimed such an idea or principle in any form or shape, but has ONLY claimed a peculiar mode of *arranging the wheels and the swivelling connection.*

HE HAS NOT stated or claimed as any part of his invention, or as any part of his principle, the *mode of traction.*

3d. HE HAS stated one mode of traction, (substitution of ordinary perch bearing carriages,) which is absolutely inconsistent with any such *alleged idea or principle.*

4th. If Plaintiff were the *real* inventor of such *new idea*, HIS PATENT should be for a COMBINATION of the parts essential to the embodiment and practical development and use of that idea.

PATENTS for IMPROVEMENTS on MACHINERY must describe the parts or COMBINATIONS.

5th. Plaintiff must SURRENDER and get a RE-ISSUE, if such is the case.

TRUE PRINCIPLES OF THE EIGHT-WHEEL CAR.

What are the true mechanical principles of the construction and operation of the eight-wheel car?

All that is matter of *elemental principle*, as distinguishing the eight from the four wheel car, is the SWIVELLING PRINCIPLE.

The essential characteristics of construction and arrangement of the eight-wheel car are :—

1st. Two trucks under one body, the body to be of sufficient length to allow the trucks to swivel without collision against each other.

2d. The trucks should be placed far enough apart advantageously to support the body upon the trucks, it being generally found in practice expedient to place the bolsters about seven or eight feet from the end of the platform or bottom framing of the car body.

3d. The trucks themselves should have four wheels each.

4th. Held in rigid wheel frames that are well braced to keep them square.

5th. The wheels in each truck should be distant apart from centre to centre or between the bearing points about the same as the gauge or distance between the rails of the track, and form a square on the track from one bearing point to the other.

6th. The middle or centre of each truck must have a king, or SWIVELLING bolt connection with the body.

7th. And the body must have side bearings in the truck frame, to steady itself and prevent rocking when in motion.

The distance between the bearing points of the wheels on the rails is the essential and elemental feature. The distance between the flanges is not essential or material; but results from the diameter of the wheels and the distance between the bearing points on the rails.

It is the action of the bearing points of the wheels upon the track, and the reaction of the track upon those bearing points, which govern and control the motion of the car upon the rails.

As to the distance of the bearing points of the wheels upon the rails in each truck, it should be about as far apart as the gauge of the track.

1st. BECAUSE bringing the wheels nearer together can never be NECESSARY or useful, since no railroad is now built with curves so sharp as 1 *inch or* 1½ *inches* in 4½ to 5 feet ; i. e., so as to use up all the play between flanges and rails.

2d. Bringing the *wheels* NEARER is followed by many unnecessary evils.

(*a.*) Both wheels are thrown up in too rapid succession by any inequality in track.

(*b.*) UNNECESSARY wrenching the rails by the re-action necessary to keep the truck in position on the rails, by LOSS OF LEVERAGE.

(*c.*) Consequent increase of tendency to fly off.

(*d.*) Extra *wear* and *tear* of flanges and rails.

(*e.*) Loss of *motive power* by INCREASE of *friction*, resulting from greater re-active force and *obliquity of truck*, or twisting of truck against the side of the rail.

(*f.*) Twisting of truck, increasing its tendency to catch at the junction of the rails, or at brakes, and consequent tendency to hop off.

The bearing points should NOT BE NEARER together than the width of the gauge of track,—i. e., the distance of centre of axes should equal the width of the track.

REASONS.

1st. The bearing points being not only equi-distant from each other, but being the four corners of an equi-lateral parallelogram, or square, of which the king-bolt is the centre, the action and re-action of the truck in any and every direction is equalized and balanced at the king-bolt, and thus the truck remains balanced and steady upon the track.

In other words, the mesne line of the action and re-action, (the resultant,) passes through the centre of the king-bolt.

If the bearing points were *nearer* together, this mesne line, or resultant, would pass *one side* of the king-bolt, CAUSING

(1.) An excessive friction of one *flange* against its rail.

(2.) Tendency to hop off, in consequence of the *angle of pressure* of flange against the rail being more *direct*.

2d REASON. (*Illustrated.*)

If a force of one pound tends to twist the truck round, then the RAIL must re-act and resist to the extent of one pound. If these forces are applied at EQUAL LEVERAGE,—equal advantage,—they will be equal. If not applied at EQUAL LEVERAGE, the want of leverage must be supplied by an increase of force.

The action of the flanges upon the rails, and the *re-action* of the rails upon the flanges, may philosophically be considered as FORCES applied upon *levers* whose fulcrum is the king-bolt. The *length of the levers* (at which those forces act, that tend to twist the truck off the track), is equal to half the width of the GAUGE.

If the bearing points are not equi-distant, the leverage (of those forces which twist the truck back into its proper position, viz., the re-action of the rail,) is LESS, and is equal to half the distance of the bearing points on one side of the truck from each other.

[Mr. Whiting then explained these ideas to the Court by aid of Models; and then proceeded:—]

The FORCES may be considered as *delivering* themselves; one part on the line of the track, and the other part on a line across the track.

Some of the CONSEQUENCES of too close proximity of the BEARING *points* and loss of leverage, are,—

1st. Unnecessary stress on the *flange* of the front wheel while passing curves.

2d. Twist of truck and danger of hopping off track.

3d. Unnecessary friction and loss of power and wear and tear of rails, &c.

4th. Throwing up both wheels nearly simultaneously.

The bearing points of the wheels should be *as far apart as the nature of the curves will possibly allow*, because the farther they are apart the *less the truck can swivel* between the rails before it shall have exhausted all the play room or lee way between the flanges and the rails; hence the wheels will always be kept more exactly in the line of their forward motion, more even and steady, and require less STRESS on the RAILS and FLANGES, less FRICTION, less LOSS OF POWER. THESE ARE THE TRUE PRINCIPLES. For a list of all our experts to this point, see "INFRINGEMENT."

OBJECTIONS TO OUR MODEL.

There were none made to it before Judge Conkling, on the former trial, and my colleague states that it was admitted to be correct in the Troy case.

The objections now made are :—

1st. That the springs are not straight, but that they are eliptical.

This is true, but they were like the springs in common use at the time, as is shown by all drawings.

There is no difference pointed out in the patent, between the straight and the curved springs.

There is no difference in the principle, and the only direction in the specification, is to have them *twice as strong* as those in common use, and short leaves on the *top* of the long ones.

All our experts say that this is what the patent describes, and the Plaintiff's witnesses do not deny it.

No model is produced by the Plaintiff, to illustrate his invention. It cannot be compared with any drawings of the original patent, if any existed, as he does not favor us with any here.

The difference of motion is extremely slight between the two forms of spring.

The Court will recollect Mr. Keller's argument to show that the centrifugal force of the cars, in passing curves, would throw the weight over upon the outward spring, and thus tend to cause the axes of the wheels more to conform to the radii of the curves, and so pass smoother along the track.

To accomplish this, the springs must be eliptic or curved; for, if the springs were straight and the ends fastened, the depression of them would bring the ends nearer together, and thus aggravate the difficulty, by bringing the axes ACROSS the radii.

The Court will perceive the morale with which the Plaintiff's case is prosecuted. He claims the *advantage peculiar* to that kind of spring, in one breath, to serve one purpose; and then, to serve another purpose, turns round and says that the model is erroneous, because it has that kind of springs!

The 2d objection is,—That the body is too narrow, and that it shows the bottom of the carriage below the tops of the wheels.

Answer. This is not material, as the trucks swivel all that is necessary.

It is only made to show the connection with the trucks. It allows of perfect swivelling, and that is all that is important.

It illustrates the peculiarities as to the arrangement of the wheels on the bearing carriages, and their connection by the king-bolt to the body.

No witness or expert objects to it, and, in fact, all our experts testify to the perniciousness of the invention from examining the PATENT itself and not the model; and if a structure is not an available one, it is no fault of ours.

3d. Then it is objected that the outside bearings are not shown in the model.

Answer. This makes no difference as to the arrangement of the wheels and their connection with the body, and it was unnecessary to show in the model that which is embodied in another patent, and which is no part of his claim in this. No person would have had a right to use them, if he had bought this patent right of Winans. Why, then, should our model incorporate that which is no part of his patent? It might as well embrace the brakes, and all the other apparatus connected with it, and with which he has no concern. The descriptions in the patent are inconsistent with allowing the side bearings on the outside of the wheels. He says, "Upon this first bolster I place another of equal strength, and connect the two together by a centre pin or bolt passing down through them, and allowing them to swivel, or turn upon each other, in the manner of the front bolster of a common road wagon." The bolsters of a common road wagon always swivel *inside* of the wheels, and the body is narrower than the distance of the wheels; and if *straight* springs were used, and these springs were bolted to *outside* bearings, the lower *bolster* would come between the flanges of the wheels, and thus prevent the wheels from being *very* close together.

We have not done the Plaintiff injustice, for both sorts of bearings are in use at this moment on the New York and Erie Railroad. See affidavits of Charles Minot, Defendants' No. 3,

p. 161, fol. 642; and John Murphy, No. 3, p. 11, fol. 43, where he says, " Eight-wheel cars with *inside* bearings and fixed boxes to the wheel frames and circular bearing and axis, are now in use on the Philadelphia and Norristown and Germantown Road;" so that both inside and outside bearings being in use at this time, the criticism is unfounded.

This very model was before the Court on the first and second hearings of this cause, and no objection has been raised to it heretofore.

No expert has objected to it—the objection is merely on the part of Counsel.

To the suggestion of Counsel, that the body is too narrow, and does not project over the wheels, I answer,

1st. It is not material, if the trucks swivel sufficiently to turn all curves.

2d. It is not the object to show any structure of body.

3d. If you use straight springs, as the Plaintiff demands, how are you to prevent the body from being below the wheels, unless you use enormous bolsters, and risk the structure breaking, in actual use, by straining the king-bolt.

It is a false charge to assert that our model was made for the purpose of distorting the proportions of the Plaintiff's invention.

WINANS'S THEORY.

The next topic to which we intend to call the attention of the Court, is the PATENTEE'S THEORY *of construction of the car, displayed in the specification; and we shall show it to be erroneous in mechanical principle, and pernicious in practice.*

The idea, that bringing the flanges of the wheels very near together, would cause the axles to coincide more nearly with the radii of the curves of the track—as a mere abstraction—is correct. It would be practically correct, if there were no play room between the flanges, and if the truck were never to be put in motion. But such is not the condition of the flanges and rails upon any railroad for public travel.

The peculiar construction of the elastic, or spring connection of the axis, allows them, while passing curves, to be *twisted* in a direction *transverse to the radii* of the curve, and so tends to cause the trucks to hop off the track.

The *same difficulty*, though to a less extent, attends the use of *any description* of truck frame, provided the axles are near together; so that these remarks are as applicable to the *common* bearing carriages used in 1834, and mentioned in Winans's patent, as a substitute for his spring trucks, as to the spring trucks themselves, since the spring trucks merely aggravate a difficulty which is inherent in the too close proximity of the wheels.

Inasmuch as lateral play is allowed between the flange and the rail, the nearer the axles are brought together the farther they may swivel in a direction across the radii; while, on the contrary, the farther apart the axles are placed on the truck, the less is the *space on a curve*, in which they can swivel between the rails; therefore the nearer do the axes at all times coincide with a perpendicular to the tangent of the curve at the bearing points. Therefore the more nearly parallel with the rails will be the line of forward motion of the wheels, and the more coincident with the radii of the curve will the axes be compelled to remain while in motion.

So true is Winans's theory on paper, and so untrue in practice, when the trucks are moving round a curve; and if he has any advantage in theory over the old-fashioned trucks of Chapman, Tredgold, or Quincy cars, or those of the Defendants, the mechanical defects of the structure he recommends, are such as to embrace all that is erroneous in the theory, without securing a single practical advantage.

The Plaintiff contends that it is best to bring the wheels as near as possible together, in each truck, because he imagines that friction is most avoided when you bring the wheels closest together, and that then the *least resistance* is offered by the *flanges to the guidance of the rails*. In other words, the nearer the truck can be made to act *like a single pair* of wheels, swivelling like a bolster of a road wagon, the better, as Winans thought.

This is all a FALLACY, founded in the mistake of carrying out a theory, that is well enough on paper, to a foolish extreme, so as to lead to error in practice.

1st. IN FACT, there is NO saving of friction in passing curves, such as now are allowed on railroads, by bringing the bearing points of the wheels nearer together than the width of the track. This has been demonstrated theoretically, and shown to be true in practice.

2d. If there is quite sufficient room for the wheels at *this distance apart* to pass along *the curve* without USING UP the play room between the rail and the flanges, there is no more friction of the flanges on the rail than if there were three or even five times the play room left.

If the flange is not compelled to *touch the rail at all*, (by reason of not occupying the play room,) that is the same in its effect, whether great or little SPARE room is left.

3d. If there be ANY PLAY room *left*, so as to allow the truck to pass AT ALL, then is LESS resistance made by the flange (*to the guidance of the rail*, in twisting the truck round,) the FARTHER APART the wheels are in the truck—because the action of the rail upon the flange is at a GREATER LEVERAGE, as above demonstrated.

4th. On the STRAIGHT track the friction is LESS, the steadier the forward MOTION of the TRUCK; and it is conceded, that the farther apart the *wheels are*, the steadier the motion, therefore the less the FRICTION on the STRAIGHT TRACK, which is 9-10ths of all railroads.

5th. But even on curves, the steadier the motion the less the friction, because the EXCESSIVE *play of the truck* brings the wheels at too great angles with the track, not only making extra friction, but danger of hopping off.

6th. The distance of the wheels is not to be adjusted SOLELY with regard to FRICTION, but other elements must be taken into view.

1st. SAFETY; that is, liability to run off the track.

2d. The action of the wheels on the rails, in wrenching them out of place.

3d. The *speed* must be considered—and the greater the speed the more DISTANT must the bearing points of the wheels be placed —(within the limits hereinbefore stated.)

4th. The friction.

5th. The span of rail or width of sleepers, and the distribution of WEIGHT on the road.

6th. The length of radii of curves, and various other elements not necessary now to be enumerated.

And it will be found, that NO *road* now exists, used for running at high speed, having curves 1½ inches in five feet of length.

And therefore it CAN NEVER be necessary to bring the axles of the wheels nearer together than the width of the track.

All the four-wheel cars and all locomotives have the large wheels much more distant (in reference to these bearing points) than the *gauge of the track*, and as THEY can pass all the curves and switches of our roads, of course other trucks with wheels still *nearer* than these, can, " *a fortiori*," do the same.

The fact that the outer front wheel in all cases MAY touch the inner side of the upper rail in passing curves, makes no difference, as this will happen, *however* near or distant you place the wheels, owing to the tendency of moving bodies to go in RIGHT LINES.

The only question is, *which is the easiest way* to turn the moving body *out of a right line* and into a *curve*,—by prying it round with a LONG LEVER, or a SHORT one ?

It is another fallacy to suppose that the trucks will run BEST, that is, with least friction and most safety, when they *have the greatest freedom of swivelling motion.*

1st. The nearer the axes, the more they resemble a *single pair* of wheels, swivelling on a bolster.

Now such a pair of wheels have too great freedom of motion, wabble about too much; in passing curves, one wheel is pressed *back* too MUCH, so that the axis *cannot coincide* with the *radius of the curve it is passing*, but crosses it the wrong way.

All that makes it conform to the curve is the leverage of the rail applied to the UPPER wheel. The length of that leverage is equal to one half of the distance between the bearing points of the wheels on the rails, and the length of leverage of the lower wheel equals the *same* distance, the fulcrum being at the king-bolt.

One could DO NOTHING with such a pair of WHEELS, without a PERCH, or FILLS, as in common wagons.

It needs the action of the rail upon the REAR wheels to GUIDE the forward wheels of the truck; and the farther off you put the *controlling influence* of the rear wheels the better, because the greater the leverage.

Winans sets up NO SUCH THEORY as his *Counsel* set up, as to the GREAT IDEA of the patent depending on the car being drawn by the body.

THIS COURT entertained no SUCH idea in delivering the opinion we have heard read.

Winans's idea was *not to add* to the FREE SWIVELLING OF THE TRUCKS, as they always SWIVELLED freely enough BEFORE, but to DIMINISH FRICTION and make the AXES of the wheels coincide with the RADII of the curves they were passing, by BRINGING the WHEELS in *each truck close together*, and spreading the trucks far apart. AND HIS purpose would be accomplished as well when the draft was by trucks as by body; NAY, even BETTER, *in case* a SINGLE car is used ALONE.

It is only when drawn by the TRUCKS that the *forward* truck is *prevented* from the very DIFFICULTY of TOO much wabbling !*

---

* A good illustration of the danger of placing the wheels too near together may be found in the trucks of locomotives. In order to preserve the wheels from collision, they are placed under the forward end of the frame, or engine. The wheels are made of small size, and placed near together, in order to accommodate them to other parts of the machinery; and to avoid *too great* LENGTH of the main springs which bear the weight of the engine. The size and position of the wheels result from a variety of compromises necessary in order to adjust the different parts to each other. If safety alone were considered, and if this truck had nothing different from the duties of a common car-truck to perform, and if the space were ample, the wheels would be placed as they are in the common car-trucks; that is to say, with axles as far apart as the gauge of the track. But as this has not been found practicable heretofore, without throwing too much weight on the forward truck, and encountering other difficulties well known to locomotive builders, they have made trucks to answer the purpose, retaining the wheels near together, but avoiding the danger of such a proximity of wheels, by an arrangement of the parts entirely different from what is claimed in Winans's patent. In Winans's, the *whole weight* is borne upon *the centre* of THE BOLSTER, (as has already been shown,) and not upon ANY *side bearings*. The locomotives have NO BOLSTERS, and bear NO WEIGHT upon the *centre of the frame*, at or near the king-bolt.

All the weight of the front end of the locomotive is borne upon SIDE BEARINGS.

WINANS'S truck is made by uniting the axles by means of long springs, bolted to boxes which rest on the journals; preventing the springs from acting except by opening or closing the wheels on one side or the other of

Such were *Winans's theories*, and so were they founded in ERROR.

Leaving the REAR TRUCK IN DANGER, unless coupled with the forward truck, and at a great DISADVANTAGE, and the *forward truck* safe only when guided and drawn by the PERCH.

And as SOON as speed of railway travelling INCREASED, the very first change necessary to be made was to DEPART still farther from Winans's theory, and SPREAD THE WHEELS *farther apart*, otherwise they could not be trusted SOLELY to the guidance of the rails :

So different was Winans's notion from what his Counsel have stated !

Now if the arrangement of wheels, and their connection with

the truck. The locomotive spring is not *connected* in any way with the axles of the wheels; but is allowed to expand or contract freely in cast iron pockets; and is confined by a saddle, so that the motion of the springs is entirely independent of the axles; and the axles are held firmly, and the wheels square upon the track by a rigid wrought iron frame. To avoid accidents, and to prevent the trucks from swivelling too far, and from too great zig-zag motion, they have four strong chains fastened from the four corners of the truck to the body of the locomotive. These chains are made of such a length that the truck cannot swivel round so far as to allow the flanges to cross the rail; and also so as to hold up the locomotive in case of breaking one of the wheels.

The action of this truck is also modified, and in some degree controlled, by the driving wheels, which are intended to bear the larger part of the weight of the locomotive. Sometimes only one of the drivers has a flange; sometimes both. In either case the drivers, being *not* in any swivelling truck, (as *Allen's* were,) but being permanently connected with the body of the engine, modify, and to some extent control the action of the truck, and require it to act very differently from what would be the case were *two swivelling trucks* to be placed under the locomotive. The motion of the forward truck is thus made more steady than it would be if left to itself,— so that the truck of a locomotive is far from being, in any respect, like that recommended by Winans for a freight or passenger car, either in its *structure, uses, controlling influences, checks and safeguards*, or *modes of operation*.

The locomotive builders are now beginning to remedy this defect, by arranging the machinery so as to allow them to place the wheels farther apart, and thus approximating the true principle of the truck, as established by Chapman, Tredgold, Allen, Bryant and others. In some of the most recent engines the bearing points of the wheels are *more* distant from each other than the *gauge of the track*. And it is found that the change causes the engine to run more steadily and smoothly upon the track.

the body is made in the manner *particularly recommended by Winans*, the car is DANGEROUS, and UNFIT FOR USE, as stated by Mr. WATERMAN.

See his affidavit, Defendants' No. 3, p. 65, and there the Court will find five specific objections to the use of the railroad truck as described by Winans, whether you use it with or without the spring.

And your Honor will see that he, in common with many other scientific experts, condemns the use of the truck, not only by reason of the peculiar structure of the springs and bolsters described in the patent; but because the trucks (which, as the Patentee says might be substituted for these,) also had the fatal defect of *placing the wheels too close together*.

The other experts show that there is an inherent difficulty in the theory of making the axes of the wheels coincide as nearly as possible with the radii of the curves.

Henry Waterman, p. 65, No. 3, says:

"*A car constructed as described by Ross Winans cannot be used with safety for general running* on a railroad at an ordinary velocity, such as aimed at by his specification, say at fifteen or twenty miles an hour. His car, as described by his specification, is too dangerous to use on a railroad, both to life and property, for the following reasons:

1st. *The metal composing the springs that connect the wheels is liable to constant depreciation of strength by its vibration*, and is so liable to break that it could not be relied upon, and such a break in that truck would certainly do great injury, and probably destroy the life of some of the passengers.

2d. *The next reason is*, that these *spring trucks will not keep square*, but lose their rectangular form by the transverse pressure, which will cause the truck to run off the track.

3d. *Another reason is*, that this *construction and nature of the truck, when attempting to run on a curve of the road*, where necessarily the outer wheels have the longest distance to traverse, and bear hardest against the side of the outer rail, *allows the axles of the wheels to get out of parallelism with the radii of the curve*, by the inner wheels forcing themselves forward of the outer wheels, twisting the truck out of rectangular shape, and necessarily run off the track. It is the worst form of truck that I have ever seen devised.

4th. *Another reason why it is dangerous* also, is that *the vibration of the springs* will allow *the axles to get out of parallelism* with each other, which adds to its dangerous character before explained.

5th. *The very close proximity* of the wheels required by the Winans specification, is also *erroneous in principle* and practice, as it allows a serpentine, unsteady and unsafe motion, both horizontally and vertically ; and the wheels follow each other in shocks on the same inequality on the rail, so quickly as to blend the shocks so much together as to become like the action of a single wheel, with a single shock, without giving the front wheels a chance to recover before the next one strikes, and causes the wheels by the concussion and the action of the springs, to hop from the rail, and thus run off from the track from this cause, or become dangerous; these inequalities are frequent at the joints of the rails. It is the worst plan for a truck, and all of these causes combined, constitute the very great danger and destructive character. I have, *in practice, proved these principles and actions as stated, to be correct.* The proper mode of constructing the trucks of eight-wheeled passenger cars, and that in general use, is to have a rigid rectangular wooden or iron wheel frame with pedestals securely attached to it, to insure a normal action of the wheels on the track, and hold the axles exactly parallel to each other, and parallel to the radii of the curves, and the wheels in each truck should be about 55 inches apart from centre to centre or between the bearing points of the wheels on the rails, that is, between 54 and 57 inches, with wheels varying from 30 to 33 and 36 inches diameter, which are the sizes in common use. This construction of truck runs with ease on both the straight and curved lines of the road, and with perfect safety, in the principles and action of the wheels, on the rails, at all rates of speed."

We produce the united testimony of *thirty-three men of science, consisting of practical car builders, civil and mechanical engineers, railroad machinists and superintendents*, and persons whose profession has led them to the most intimate acquaintance with this subject. With ONE VOICE they say that the railroad car, as arranged and described by Winans, *is not practical or useful*, but erroneous in the principles of its construction; and dangerous in practice. I will give their names:

The Plaintiff produces only Elgar and Hibbard on this POINT, neither having any PRACTICAL KNOWLEDGE on this subject. *Thirty-three witnesses against two!*

If it be said that some of these witnesses have formerly stated, as Mr. LEE has, that many cars, built according to Winans's patent, were in use in all parts of the country, and that this shows that they are in error, we reply that they were led into that error by the shrewd manner in which the Plaintiff states his claim, which was supposed by them to cover the SWIVELLING PRINCIPLE.

The witnesses on the part of the Defendants, to whom I have referred, are:

1.	Henry Waterman,	Defendants' No. 3, p.	65.
2.	William T. Ragland,	"	"	67.
3.	George Stark,	"	"	70.
4.	George W. Smith,	"	"	82.
5.	Charles B. Stuart,	"	"	90.
6.	William J. McAlpine,	"	"	93.
7.	Vincent Blackburn,	"	"	96—7.
8.	Walter McQueen,	"	"	103.
9.	George S. Griggs,	"	"	110.
10.	William Raymond Lee,	"	"	111.
11.	Albert S. Adams,	"	"	114.
12.	James H. Anderson,	"	"	118.
13.	Jno. B. Winslow,	"	"	121.
14.	Asahel Durgan,	"	"	124.
15.	Jno. Crombie,	"	"	126.
16.	H. W. Farley,	"	"	129.
17.	Robert Higham,	"	"	133.
18.	Albert Bridges,	"	"	135.
19.	Jeremiah Van Rensselaer,	"	"	137.
20.	Isaac Adams,	"	"	139.
21.	W. C. Young,	"	"	150.
22.	W. P. Parrott,	"	"	153—4.

23.	M. W. Baldwin,	Defendants' No. 4, p.		4.
24.	Richard French,	"	"	5.
25.	Henry Shultz,	"	"	44.
26.	Wm. B. Aitken,	"	" 3,	31.
27.	Jacob Shryack,	"	" "	40.
28.	Godfrey B. King,	"	" 4,	81.
29.	Stephen Ustick,			
30.	Oliver Byrne,			
31.	Edward Martin,			
32.	Stephen W. Worden,	"	" 2,	39.
33.	Henry Shultz.	"	" 3,	44.

Now there are the names of thirty-three men of science, and of practical knowledge of railroads, who inform this Court, that the invention as described in Winans's patent is of no practical utility, but dangerous and pernicious.

I ask the gentlemen on the opposite side to find a single car, built in actual conformity to Winans's plan, in existence. I know they tell us that his cars are scattered all over the country, but this assertion *assumes the point in controversy*. There is undoubtedly a vague presumption and hollow pretence by the Plaintiff, that many and most of the cars in the United States are on his plan; but where is one to be found in use, actually embodying his *peculiar devices?* Not one car has been proved to have been built, whose dimensions, proportions and parts really correspond with the patent.

In fact, the car of Winans's was actually tried on the Baltimore and Ohio Railroad and CONDEMNED. See

Jacob Shryack,	Defendants' No. 3,		p. 40.
Stephen W. Worden,	"	" 2,	39.
Henry Shultz,	"	" 3,	44.

No railroad has now in use trucks under the cars, with wheels having their flanges "as near as may be without touching." No such principle is allowed; we produce an overwhelming weight of testimony to the contrary, viz.:—

1st. See Walter McQueen; Defendants' No. 3, p. 100, for the fact *that the Defendants have the axes of the wheels at a distance apart about equal to the breadth of the track.*

2d. Septimus Norris, one of the most eminent car-builders in any part of the country, Defendants' No 3, p. 4.

3d. Richard French, Defendants' No. 3, p. 5, and who altered the car "Victory."

4th. Henry Shultz, Defendants' No. 3, p. 44, who places them as far apart as he can get them in a truck which is from seven and a half to eight feet long, on the Philadelphia, Wilmington and Baltimore Railroad.

5th. Waldo Higginson, Defendants' No. 3, p. 83, an engineer on the Boston and Lowell Road, an educated gentleman whom I know personally to be a man of high character and of accurate and practical mind.

6th. W. T. Ragland, Defendants' No. 3, p. 68, who states that the distances at which he places the axes apart is 54, 55, and 56 inches, in bearing trucks, and *which distances, he says, are generally adopted*.

7th. George Stark, Defendants' No. 3, p. 70.

8th. Vincent Blackburn, Defendants' No. 3, p. 96.

9th. George S. Griggs, Defendants' No. 3, p. 104, who places the distance between the flanges at twenty-seven inches.

10th. Wm. R. Lee, Defendants' No. 3, p. 110.

11th. Albert S. Adams, Defendants' No. 3, p. 114, who states that the best arrangement is, to place them as far distant as the square of the track.

12th. J. H. Anderson, Defendants' No. 3, p. 117, who says that on the Providence and Stonington Road, they are placed apart a little more than the breadth of the track.

13th. J. B. Winslow, Defendants' No. 3, p. 20, who states that the axes are as far apart as the gauge of the track.

14th & 15th. Asahal Durgan and Henry W. Farley, Defendants' No. 3, pp. 123 and 129, state that the bearing points of the wheels should be as far apart as the gauge of the track.

16th. John Crombie, Defendants' No. 3, p. 125.

17th. Davenport and Bridges, Defendants' No. 3, p. 135.

18th. George Beach, Defendants' No. 2, p. 56.

19th. L. R. Sargent, who says that the flanges are generally from twenty to twenty-seven inches apart, Defendants' No. 2, p. 41.

Now who can say, (after we have brought the most eminent engineers in the country here, unless they are, as Mr. Keller has boldly charged, a set of perjured men,) that there is any pretence that the modern railroad cars incorporate the principle of bringing the axes of the wheels as close together as may be, to make them coincide with the radii of the curves?

From the foregoing it clearly appears:—

1st. That the Plaintiff was not the first and original inventor of *the arrangement of the running gear* of the eight-wheel cars as *built by the Defendants*, nor of any part or combination thereof, and this results from our examination of prior inventions.

2d. That the eight-wheel car grew gradually out of the necessities of railroads in their earliest periods of construction, and was suggested to, and built by those, who were earlier on the ground than Winans, to whom the railroads owe absolutely nothing.

3d. That Winans was not the inventor of the Columbus; but that whether he was, or was not, if the car itself ever existed, Winans's patent is void.

4th. That Winans was not the first inventor, nor *an* inventor of any new principle, applied in any way, but borrowed the philosophy of his specification from Tredgold, Allen, Jervis and others.

5th. We have shown what in reality are the true principles of the eight-wheel car; and,

6th. The erroneous principles adopted by Winans in his theory, and their utter failure in practice.

7th. I will merely add, that Winans seems to have drawn his specification so skilfully as to have entirely misled the public as to the real nature and extent of his claims.

INFRINGEMENT.

Having now shown what Winans's claim really is, so far as the patent can be considered valid and the extension legal, (*See Construction*, p. 40,) I now proceed to show that *the Defendants have not infringed upon the Plaintiff's rights, whatever construction shall be given to the Letters Patent.*

There is, it is true, one construction upon which all the world would be infringers; that is, if Winans has the *exclusive* right to build and use every sort of eight-wheel cars, with *two swivelling trucks*:—and this construction, which is now admitted to be erroneous, was at one time supposed to be the true one by some of the Defendants' experts, who, upon THAT mistaken hypothesis, stated that the cars in general use were substantially the same as described by Winans's patent.

But as this construction would include the Chapman, Tredgold, Quincy, Baltimore Timber cars, Wood cars, Trussell cars, Allen's steam carriage, Columbus, Fairlamb, Jervis, the Dromedary, Winchester, Comet and Victory, &c., &c., the patent would be void, as embracing claims far broader than the invention; and the Plaintiff having filed no disclaimer, could not recover at law, nor have any injunction. (See Chief Justice Taney's opinion, Plaintiff's proofs, p. 21, Art. 3.)

Under any other reasonable construction the Defendants do not infringe.

What proof of infringement has the Plaintiff offered? On whom does it rest? What disinterested party has sworn it? Who has averred that the cars manufactured by the Defendants are substantially the same thing as the patent describes?

No man has said it, except CHARLES D. GOULD.

Is Mr. Gould a disinterested party?

Is he an expert, or is his opinion to weigh against that of a large number of the most scientific and practical men in the country?

How do the Plaintiff's witnesses discriminate between the Winans arrangement and the older cars? What do they rely upon, as showing a substantial difference, so as to maintain Winans on the question of novelty?

The Plaintiff's counsel say that the difference consists in the drawing by the body. If this be all that distinguishes the Plaintiff's from prior inventions, then the Plaintiff's case fails, because the patent makes no such claim. But, if his patent cover the things described in his claim, and no more, the evidence is conclusive that we do not infringe; and the evidence is also conclusive, that whatever the Plaintiff claims, the *Defendants use only that which was known and used prior to Winans.*

The testimony of the following witnesses clearly proves *that* WINANS'S *" mode of arranging the wheels, and connecting them with the body," is impracticable and dangerous; and is entirely and essentially different from that which is used by Defendants,* and which is in general use in the country.

In other words, there is no infringement of the Plaintiff's claim.

1.	Charles B. Stuart,	Defendants'	No. 3,	p.	90.
2.	George W. Smith,	"	"	"	82.
3.	W. T. Ragland,	"	"	"	68.
4.	Henry Waterman,	"	"	"	65.
5.	Jacob Shryack,	"	"	"	40.
6.	William J. McAlpine,	"	"	"	93.
7.	Vincent Blackburn,	"	"	"	96–7.
8.	Walter McQueen,	"	"	"	103.
9.	George S. Griggs,	"	"	"	110.
10.	Albert S. Adams,	"	"	"	114.
11.	James H. Anderson,	"	"	"	118.
12.	John B. Winslow,	"	"	"	121.
13.	Asahel Durgan,	"	"	"	124.
14.	John Crombie,	"	"	"	126.
15.	Henry W. Farley,	"	"	"	129.
16.	Robert Higham, (2d affidavit,)	"	"	"	133.
17.	Albert Bridges, (2d affidavit,)	"	"	"	135.
18.	Jeremiah Van Rensselaer,	"	"	136–7.	
19.	Isaac Adams, (2d affidavit,)	"	"	"	139.
20.	William C. Young,	"	"	"	150.
21.	William P. Parrott,	"	"	153–4.	
22.	Richard French,	"	No. 4,	"	6.

I pause for a moment in this argument, as I recall to mind an attack made by the able counsel who opened the cause, upon five gentlemen, of whom some are known personally to me.

It has been asserted that " they held their oaths so loosely, that they can one day come into Court and swear one thing, and on the next day another."

In explanation of the testimony of these witnesses, it is only necessary to say to your Honor, that in their first affidavits they have shown that they supposed Winans's claim covered all swivelling cars, and that the Defendants' cars (being swivelling cars,) were substantially like those claimed by the Plaintiff, merely because they were SWIVELLING cars, and for no other reason.

Afterwards, in a further examination of the patent, and, perhaps, learning from the opinion of the Judges of the Supreme Court of the United States, that Winans's claim did not cover *every description* of *swivelling* car, they stated that they were mistaken in their first affidavits in supposing that Winans's claim covered ALL SWIVELLING TRUCKS; and perceiving this error, they at once saw that the Defendants were not infringers upon the Plaintiff, because they did not USE THAT which was claimed by Winans, as distinguishing his improvement from what was known and in use before his day.

And it is a painful thing to one who entertains the sentiments which I cherish towards some of those gentlemen, to listen in silence, at this distance from their home and from mine, to so extraordinary and so bitter an attack upon their character, made only because they have seen fit to do justice to your Honor, to themselves and to the Defendants, by correcting an error into which they were inadvertently led by the loose and artful terms in which the patent of the Plaintiff was couched. But I should be no less forgetful of their dignity than of that respect which is due to this Court and to myself, if I condescend to defend them from such a charge.

Mr. Keller says, " that the affidavits all state that the Tredgold, Chapman, and other cars were the *same* as Winans's; and yet that the Defendants do not infringe."

This, Sir, is a most erroneous conception of the testimony.

The Defendants' experts swear that Tredgold, Chapman, Allen, Quincy and other cars contain the GENERAL construction and mode of operation, (that is, the principle) of the eight-wheel cars now in common use; that so far as Winans embodies the same principles, they are all alike, and substantially the same as Winans's.

That *that which distinguishes* Winans's from these preceding inventions, is *not* found in the cars in common use, nor in those built by the Defendants, but they have availed themselves only of what was well known prior to the patent, with such changes of proportion and improvements as they had a right to adopt, not one of which was invented or claimed by the patentee.

Mr. Keller, by his observation, shows that he had wholly misapprehended the true nature of our defence.

Proceeding with the argument, I ask, what expert in the United States has come forward, and stated that the cars, made and used by the Defendants, are in fact upon the same principle, and substantially the same car as is described in the Plaintiff's patent?

I answer—that excepting Gould, not one! and, what is more singular, even Winans himself has not said so. He does not dare to come into this court and make that allegation upon the record.

He has made his bill of complaint, and sworn to it.

He has made his affidavit, which is a long one and embraces many topics, and has sworn to it. He has had the aid of Counsel not surpassed in ability by any that the country affords; and, Sir,—with their advice, perhaps, or without it—he has not ventured to come before your Honor, on his solemn oath, and state that the Defendants, in their cars, embody one particle of the invention of which he was the author.

And it is one of the most remarkable circumstances that I have ever known in a patent case, that a party should come into a Court of Equity, and attempt to prove an infringement by such testimony as has here been presented; the Plaintiff himself daring not to allege it, and finding but one gentleman—(proved to be interested in this suit and in the patent, and who ought therefore to be a party to the record)—who states that cars, *substantially* like the Plaintiff's, have been built by the Defendants. And Mr. Gould, after all, only asserts what is *his opinion;* and Mr. Winans only swears that he has been *informed by Mr. Gould* that Defendants' cars are substantially like what are described in his patent! Has not Winans seen Defendants' cars? Then why does he not swear of his own knowledge, instead of swearing upon the information of Gould, who is no mechanical expert, and not legally qualified to pronounce an opinion on any such question in a court of justice?

These two gentlemen's opinions are to be put into the balance against those of many disinterested practical engineers.

Now, Sir, if you choose to examine for yourself, trusting no person, either on one side or the other; if you take the model of the car built by the Defendants, and set it by the side of a true and just model of the car built by Winans, I ask your Honor to tell me whether they are substantially the same, or not, in *those parts in which Winans differs from those who went before him?* Examine the car made by the Defendants, as to its proportions, arrangement and details of construction, and make the comparison.

The length of the body is generally now about fifty to sixty feet, five times to six times as long as the old four-wheel car, and twice or thrice as long as Winans recommends.

The distance of the wheels in each truck, or of the flanges, apart, is from twenty to twenty-five inches. (See George Beach's affidavit, p. 38, Defendants' No. 2.)

On the Hudson River Railroad, the average speed is from thirty-five to forty miles per hour, and in the cars that run on that road, the flanges are far from being " as near as may be" without touching. They are farther apart than they were in the trucks mentioned as in common use in the patent itself, viz., three and a half feet from centre to centre, and this was a greater distance than had frequently been used on the Baltimore and Ohio Railroad before 1830. Now three and a half feet from centre to centre of the axles, is forty-two inches; and the size of the wheel being at least eighteen inches in diameter, would bring the flanges within six inches of each other on that class of trucks which were built with wheels of these dimensions; yet Winans proposes to bring them still nearer. The *Defendants never brought the flanges so near* as they were in these old-fashioned bearing carriages, but they place them about *two feet apart*, and always far enough apart to admit of a brake being placed between them.

As to the springs; the Plaintiff admits, that they were in use before his invention, and he does not claim them. He states that he prefers his long spring, twice as strong as the common spring on the four-wheel car, to the old way of giving one spring to each wheel. The Defendants prefer the old-fashioned way of applying one spring to each wheel in both trucks, as Mr. Allen did in 1830, as shown in his drawings and model; and as was done by many others.

The Plaintiff prefers to have the bolster swivel on a centre pin, or king-bolt, like the common road-wagon, without side-bearings; but the Defendants prefer side-bearings, as in the Quincy car and Allen engine, &c., and a transom-plate, being Imlay's patent, or what is commonly called so.

The Plaintiff prefers to fasten his body on to a bolster, and abandons the swinging thorough-braces of the old Columbus, and trusts for his ease of motion solely to the elasticity of those springs, which, by their action, will throw the truck off the track. The Defendants prefer the swinging bolster, which is an admirable way to provide for easy lateral as well as perpendicular motion.

The Plaintiff, as before stated, gives no direction or description as to how the car is to be drawn, whether by the perch or by the body. His drawing, annexed to the patent in 1837 or 1838, is out of the question. He makes no claim to any mode of traction, and he has no reason for making any. The Defendants draw by the body, but they also have an apparatus called a spring-coupling, which does not draw by the ENDS OF THE BODY, but from a fixed point some feet distant therefrom, in order that the line of traction may more nearly coincide with the line of the track. This answers Mr. Keller's objections to coupling the ENDS together.

The Plaintiff prefers connecting the axles of his wheels by a spring twice as strong as usual in the four-wheel car, because the body is twice as long, and is to carry twice the load. The Defendants prefer the old-fashioned rigid rectangular wheel-frame, as the mode of connecting their wheels together on trucks.

The Plaintiff prefers to have the bearing points of the wheels in each truck as near as possible, and far less distant than the gauge of the track. The Defendants prefer them usually of the width of the track; while on the swiftest trains and on roads with the smallest curves, they make them as wide as the curves will allow, which is sometimes six feet apart, after the old fashioned cars of Chapman, Tredgold, Allen, Quincy car, &c.

The Plaintiff prefers that the action of each truck shall, as nearly as possible, resemble the action of a single pair of wheels, by means of the approximate coincidence of the axles with the radii of the curves of the track. The Defendants try to get as far as possible from that, and to separate the axles as far apart as may be consistently with preserving sufficient play room between the flanges and the rails when passing curves.

The Plaintiff prefers placing his truck *at the ends* of the body,—as in the Quincy car,—(or beyond it, as nobody ever did but himself,) or near the end, and never farther under the end than is necessary to protect the wheels from collision, (as in the Columbus, in which the centres of the trucks were *five feet* from the ends of the body.) But the Defendants prefer putting their trucks at least *seven feet* from the ends of the platform.

If the Plaintiff's claim is for a combination of old parts, which he says is to produce an old or new effect, what are the parts he claims to combine?

Are they the body, the trucks, the wheels, the springs, &c.?

All these have been combined in every eight-wheel double truck swivelling car before.

If the Plaintiff claims that combination of these parts which distinguishes *his* arrangement from all others, then what are the parts or elements of his combination, considered as a combination? They are

1st. The body.

2d. A swivelling truck of peculiar structure, with long springs.

3d. A wagon bolster, bearing the load on its CENTRE, and with no side bearings.

The Defendants use no such combination. What is *peculiar to Winans* they never use; but they fall back upon the old and superior combination, which Winans in vain endeavored to improve upon.

If the Plaintiff says that his claim is for connecting the body with the trucks, by a *swivelling pin or king-bolt*, then his patent is void, because he did not invent it and does *not* claim it.

If he claim a patent for putting the flanges nearer together than before, and making the body about twice as long as the four-wheel cars, whereby the trucks go farther apart, the Defendants reply,

1st. That the Plaintiff has brought his wheels no nearer than they were in the OLD PRECEDING cars.

2d. That he made his body no longer than the Trussell cars, the Wood cars, &c.

3d. That if he did this, it is a mere change of proportion, as the flanges approach nearer by merely increasing the size of the wheels.

4th. That the Defendants do not employ either of the Plaintiff's proportions, but *place their flanges as far apart, and even farther, than in the old cars*, and their body is *five or six times as long* as the old four-wheel cars; and that if such change of proportions gives any right to a patent, the Defendants have as good a right to a patent for cars sixty feet long, as Winans has for a car twenty feet long.

That if Winans's claim is construed as we have already shown it should be, (see p. 40,) there is no pretence that we have infringed either the wagon bolster, or the spring truck, or that we have stolen or applied the philosophy,—shown to be false,—of making any truck that will act like a single pair of wheels, or that the Defendants have fallen into the error of adopting Winans's theory of attempting to make the axes of the wheels coincide as nearly as possible with the radii of the curves of the road, by the near proximity of the wheels.

The Defendants have not abandoned the safe and well-tried mode of connecting the trucks with the body by means of the old-fashioned king-bolt, or transom, with side bearings, and have not approved of or adopted the Plaintiff's wagon bolster, with the weight of the car borne upon the CENTRE thereof.

The Defendants have not adopted the rickety, twisting, dangerous *spring truck* of Winans, but have continued the use of the old solid square wheel frames; and instead of trusting the safety of their truck to the strength of a long single spring, and using that as one side of the wheel-frame, which, in case of breaking, will inevitably let the car body on to the wheels, and thus cause destruction, the Defendants have used a solid wheel frame, independent of and in addition to the springs; and so that in case these springs break, no damage to the body need follow, and so that the wheels may be kept *square on the track*.

The Defendants have not adopted the near proximity of the wheels in each truck; but have kept the axes equi-distant with the width of the track, as did Tredgold, Chapman, Quincy cars and others. They have indeed put their trucks under the body a few feet from the end, as there was no other place where they could be put in order to support the body of the car; but they have not placed the trucks so near the ends of the body as the Quincy car.

If, then, (as we have shown before,) Winans has discovered nothing new *in mechanical principle*, nor any thing new in *mechanical philosophy*, but has only arranged the same parts before known, in a manner very similar, if not exactly the same as had been before done, his patent can only be for that arrangement which is substantially different from that which had been before made and *that which is peculiar to him, in contra-distinction from what was used before*. In considering what is peculiar to him, we must take the whole machine together as an organization. The connection of the wheels by springs, the near approach of the axles, so as to make them approximate the action of one wheel, the suspension of the load in the centre of the bolster without side bearings, *are all elements essential* in themselves *and peculiar* to him, introducing a great difference into the action of the truck frame from what had before been found practically useful, and from what is now used by the Defendants or by any one else.

And *the near juxta-position of the wheels* in one truck, and the *remote coupling of the trucks, are not alone considered the embodiment of what is peculiar to Winans*. These features, *alone considered*, are *mere changes of proportion*, and, therefore, not patentable. No one violates his patent who does not *combine all that is substantial* in the organization which is peculiar to Winans, even if Winans be the first inventor of it. Therefore, if the Defendants did *place or arrange their wheels as the Plaintiff placed or arranged them*, (which they do not do,) *this would be no infringement*, because this would not be taking all the essential parts, devices or elements of Winans's arrangement or combination *as he claims them;* since he does *not claim* the arrangement of the eight-wheels with respect *merely to their position relatively to each other*, but with reference to the *means by which that ar-*

rangement is carried into effect; he does not claim the connection of the truck with the body by a swivelling king-bolt, *abstractly, but with reference to the peculiar devices by which that connection is produced.*

If Winans has attempted to secure a patent for all modes of carrying freight or passengers smoothly and evenly over the curves and inequalities of the road, by means of the eight-wheel swivelling car, he has attempted to patent a result,—not a machine,—and such a patent could not be sustained for the following reasons:—

1st. It is against settled principles of the patent law to patent a RESULT.

2d. The result is not new, even if he could accomplish that result better than others.

3d. The Defendants accomplish it infinitely better than Winans,—if Winans ever accomplished any,—so that the Defendants are as far beyond Winans, as Winans was beyond the four-wheel car.

4. Mere difference in degrees, consequent upon change of proportions, is not the subject of a patent.

If Winans claim that he discovered a new principle in mechanics, and embodied it in some form in his eight-wheel car, and therefore claims a patent for his car, and all others involving the same principles—we reply,

1st. That he has discovered no principle that was not already known, printed, and practically embodied in all the swivelling cars.

2d. That he has not explained, or *claimed* in his patent, to be the discoverer of any new principle, and there are no words to that effect therein.

3d. That mere abstract principle could not be claimed and secured.

4th. That if machines are within his principle only when they can be clearly distinguished from the Quincy and Allen, wood and trussell cars, Chapman, Tredgold, &c., then the Defendants have not infringed, because Winans is on one side of all of them, and the Defendants are on the other; and the Plaintiff cannot claim novelty, without admitting that the Defendants do not infringe.

In fact, the Defendants may be said to use the Allen steam carriage, *with a box for passengers;* or the Chapman or the Tredgold, *with large wheels and a longer body,* while the bearing points of the wheels remain the same, or the Wood and Trussell cars, *with a box for passengers,* if passengers are required to be carried; but the wheels in each truck are not so near together, and the two trucks not so near the ends of the body,—

Or the Defendants may also be said to use the Quincy car, with a *box for passengers,* if they require to be carried, and with the *wheels larger,* and *the body longer.*

I have before mentioned that Winans's car was for freight as well as for passengers, as the specification and drawing show.

And as the *only real principle which distinguishes the eight-wheel from the four-wheel car, is the swivelling principle,* that being attained, all other arrangements *are mere alterations of peculiar devices, or parts or proportions.* And as the Defendants do not use Winans's peculiar devices, parts or proportions, it is self-evident that they do not infringe.

Mr. Keller admits, that the Defendants do not infringe, if they combine in their car every thing else described by the Patentee, except drawing by the body; and that if they were to draw by the perch, there would be no infringement. Id est: *we may use every thing described and claimed in the patent and drawing, without infringing,* and we become infringers only when we use that which is neither described nor claimed in the patent!

The Plaintiff wants an injunction to stop all the railroads in the country from using something which is not thus described or claimed; and which, we have shown, was no more the invention of Winans than the old Chapman, Tredgold and Quincy cars, whose principles of construction and operation have been so closely followed by the Defendants.

Is not this ridiculous in the extreme; and is it not trifling with courts of justice?

It is the settled rule of this Court to refuse to interfere by injunction, where the right of the complainant is in any degree doubtful or disputable; on which point I refer to the following cases:

Hart vs. Albany, 3 Paige, 213.
Steamboat vs. Livingston, 3 Cow. 713.
Collard vs. Allison, 4 ⸺ & Cresswell, 487.
Same case in Hindmarch, 315.
Boothe vs. Garely, 6 vol. Leg. Obs. 99.
Isaacs vs. Cooper, 4 Wash. C. C. R. 259.
Molly vs. Dowman, do. p. 1–14.
Barnwell vs. Malcomb, 3 Mylne & Cr. 738.

Now can the Court hesitate for a moment to decide that the rights of the Plaintiff are extremely doubtful, to say the least?

Sir, I consider this trial to be a desperate effort to compromit the judgment of this Court, in order to pave the way for an Injunction against the Railroad Companies of New York. The unusual number of the Counsel on the other side and their extraordinary zeal, show that such is the fact. Will this Court take upon itself the task of finally settling upon *ex-parte affidavits*, such questions as these—Whether or not Ross Winans allowed what he claims as his invention to go into public use before his patent?

Whether he has not led the Defendants to appropriate what he claims as his invention to their use, IF THEY HAVE DONE SO, by *his own conduct?*

Whether the Wood and Trussell and the other cars, that I have mentioned so frequently, existed before the Columbus; and whether the Plaintiff or Gatch was the inventor of the Columbus?

Whether the thirty-three gentlemen, connected by their profession with the actual operation of railroads, who swear that, according to their experience, the invention claimed by Winans is IMPRACTICABLE and DANGEROUS, and thus worse than useless, are mistaken?

Whether the Quincy car, the Chapman or the Tredgold cars, do or do not embody all that the Defandants use?

Whether the thirty-three men of scientific and practical knowledge, (*standing at the head of their profession,*) who deny that there is any infringement, are right; or is Hibbard (the only witness against them) mistaken?

Whether the *prior inventions*, now for the first time proved in Court, do or do not embody *all that is essential* in the organization of the eight-wheel car, as made by the Defendants, excepting those *devices* which are *peculiar to Winans*, and which *are not used by the Defendants?*

We feel confident that your Honor will not, upon mere "*ex parte*" testimony, and without the aid of a thorough trial, feel disposed to decide such questions of fact against us; that you will not feel willing, upon *questions* of *mechanical science*, to pronounce the opinions of the ablest engineers and practical mechanics of the country erroneous; that you will not deny us the privilege of having the merits of the defence, now for the first time brought before the Court,—once at least passed upon by a Jury, if the Gentlemen on the opposite side still insist upon going further with this case. We think that your Honor, in the exercise of that sound judgment and discretion which are the peculiar attributes of a Court of Equity, will not allow an Injunction to go forth, to ruin the business and blast the prospects of the Defendants under such circumstances of injustice,—(an injunction,) which cannot be really beneficial to the Plaintiff, since the true merits of the case have never yet, been heard by a Jury of the country, or investigated by any Judicial tribunal. It is true that this case has, in part, been heard and adjudged; and upon that which has been heard, the decision has been in our favor. But when our opponents appear before your Honor and undertake to overcome the judgment of this Court by new efforts, we have shown that those efforts are worse than useless, since they have caused us to make further investigation, to develop new evidence, and to produce new testimony, which we think is destined to settle the fate of this controversy forever.

Sir, it is now the third time that the Plaintiff's have come before this tribunal, and we cannot but indulge the hope that the decision in this hearing will be such, that it will be the last time they will attempt to harass the Respondents.

And, in conclusion, I have only to thank your Honor most sincerely for the unfailing patience and constant attention with which you have listened to the evidence and the arguments we have offered.

TABLE OF CONTENTS.

INJUNCTION, ... 1
  Public interests affected by this case,.. 1
  Grounds for asking for, ... 4
  Principles of Equity Law as to, ... 5
  Alleged licenses, and their effect, ... 8
  Former verdicts and judgments, and their effect, ... 10
  Defence founded on new evidence, ... 13
  Twice denied in this case by this Court, 18
  Acquiescence of Plaintiff estops him from relief, ... 20
  Alleged notice by Plaintiff to Defendants denied, ... 22
  Laches of Plaintiff after issue of Patent, ... 25
  Exclusive possession not had by Plaintiff, ... 25
  Reasons why not important to Plaintiff, and yet disastrous to Defendants,.. 26

EXTENSION OF THE PATENT, ... 27
  Not valid, 1. Because the Commissioner did not order legal notice, ... 27
  2. The notice ordered was not published, ... 28
  3. The Patent was extended, not by the Commissioner, but by the Chief Clerk, ... 29

CONSTRUCTION OF CLAIMS IN THE PATENT, ... 30
  Admissions of Patentee, ... 30
  What he states *his peculiarities* are, ... 32
  What experts suppose his invention was, ... 33
  What counsel allege it was, ...
  Objects of the improvements as stated in the Patent, ... 34
  Claim of the Patentee, (what it is NOT for,) 35
  Claim not for a RESULT (of travelling smoothly, &c.,) 36
  If the *result* aimed at by Plaintiff were attained by prior inventions, consequences of this, ... 35
  Claim *not* for *change* of *proportions*, ...36, 96

Claim *not* for bringing wheels near together in each truck, nor placing the trucks far apart, ... 36, 159
Claim *not* for connecting body and truck by swivelling bolts, &c.,.... 38
Theory of Winans as to the proximity of wheels, as shown in the language of the Patent, ... 37
Theory of Winans as to the connection of the two trucks with the body, ... 38
The true construction of the claims, ... 40
Arguments and conclusions thereon,..40–47

DISCRIMINATION, ... 47
  Questions here are different from what have been heretofore decided, ... 48
  Drawing annexed to the Patent was not restored according to statute, ... 49
  Drawing in many respects at variance with the specification, ... 50

ABANDONMENT, ... 52
  General rule, ... 52
  Many eight-wheel cars in public use before the Patent, with Plaintiff's knowledge and consent, ... 53
  Experiments alleged to have been tried on said cars; their history and results, ... 54
  Whether Plaintiff was or was not experimenting on said cars, *not material*, 60
  Witnesses to abandonment, ... 61

LACHES after issue of Patent, ... 62

NOVELTY of the invention, ... 64
  Subdivisions of the subject, ... 64
  Conduct of Winans, ... 65
  1. CHAPMAN'S PATENT, ... 66
  2. TREDGOLD CAR, ... 68
  Witnesses, ... 68
  Object to be accomplished, and means of doing it, ... 69
  Extracts from Tredgold, ... 76
  Equi-distance of wheels along the track no part of Tredgold's plan, ...72–81

TABLE OF CONTENTS.

MODE OF DRAWING THE CAR,	81
Not invented by the Plaintiff,	
No part of the Plaintiff's claim,	88
The car described by Patent drew by the *perch* and not by the body,	85
WASHINGTON CARS,	89
Not devised prior to Plaintiff's Patent, and not a result to which Plaintiff had arrived before his specification was drawn,	89
Old machine put to *new use* not patentable,	94
Illustrations,	95
Change of proportions not patentable.	36 & 96
Application to this case,	96
3. THE QUINCY CAR,	98
Its construction,	98
Witnesses,	99
Experts,	100
Objections to it,	102
4. THE ALLEN CARRIAGE,	107
Its priority to Winans,	107
Description of it,	108
Witnesses,	109
Criticisms,	110
Elements of its structure,	111
5. THE BALTIMORE CARS,	112
Construction of,	112
Plaintiff's witnesses,	114
Defendants' witnesses,	116
The Columbus, (who made it?)	117
Burden of proof on Plaintiff,	121
6. JERVIS ENGINE,	121
7. FAIRLAMB'S PATENT,	122
8. THE CAR, VICTORY,	123
Chronological order of these inventions,	124
STATE OF THE ART of building eight-wheel cars in 1830,	125
HISTORY OF THE PROGRESS OF THE INVENTION,	127
NO NEW PRINCIPLE was discovered by Winans,	131
TRUE PRINCIPLES of the eight-wheel car,	134
MODEL—objections to,	137
WINANS'S THEORY—its errors shown,	139
Locomotive trucks—(Note upon,)	143
Reasons why Winans's theory is dangerous when put in practice,	145
Witnesses to prove the Winans truck impracticable and dangerous,	147, 152
Winans's car was *condemned* as worthless,	148
Winans's cars not in use on the railroads,	148
Witnesses,	149
Conclusions,	150
INFRINGEMENT,	150
Witnesses to show that Winans's car is impracticable and dangerous,	147, 152
Error of Defendants' experts and explanation of the same,	153
Only one witness swears to infringement,	154
Plaintiff does not swear to it,	154
Difference between Plaintiff's car and Defendants', as to proportions, &c.,	155
Various suppositions as to Plaintiff's claims, and argument as to infringement upon these hypotheses,	157
Admission of Plaintiff's counsel,	161
Court will not interfere when rights of Plaintiff are doubtful,	161
Authorities,	162
Court will not decide a variety of new questions without aid of jury,	162

ERRATA.

In the hurry of passing through the press, several errors have crept in. We indicate the most important.

On the title page, instead of patent for "THE" eight-wheel car, read "AN" eight-wheel car.

In a part of the edition, p. 27, 9th line from the bottom, for "*invalid*," read "valid;" also, p. 28, 10th line from the bottom, for "patent," read "extension."
p. 31, line 3d from the bottom, after "other," insert "are."
p. 33, line 20, for "inventor," read "patentee."
p. 42, line 25, after 8 wheel, insert "cars;" line 29, transpose "the" before "wood."
p. 43, line 6, dele "alone."
p. 44, line 2, after bearings, insert "in;" line 21, for Plaintiff, read "Plaintiff's invention."
p. 71, line 17, for "whether the body be short; so that the eight wheels need not be long or equi-distant," read "whether the body *be short or long*," &c.
p. 72, line 17, dele "it."
p. 87, line 24, dele "he."
p. 89, line 18, for "Plaintiff," read "Defendants' rely."
p. 111, line 1, for "Baltimore cars," read "Allen cars."
p. 130, line 2, dele period and capital T.

Printed in Dunstable, United Kingdom